The Mathematics of Large Language Models

From Tokens to Transformers, Training to Decoding

Team Trex

Published by NobleTrex Press

For permissions and other inquiries, write to:

P.O. Box 3132, Framingham, MA 01701, USA

Contents

Introduction

Large Language Models (LLMs) represent a convergence of linear algebra, probability theory, and large-scale optimization. While often discussed in terms of their emergent capabilities, fundamentally, these models are rigorous mathematical functions designed to estimate the conditional probability distribution of a token sequence. This book provides a systematic examination of the mathematical operations that define this architecture, stripping away abstraction to reveal the explicit arithmetic and algebra underlying modern natural language processing.

The text begins by establishing the foundational definition of language modeling: the factorization of joint probability distributions into products of conditional probabilities, governed by the autoregressive assumption. We examine the transition from the discrete nature of human language to the continuous, differentiable manifolds required for computation. This involves a formal treatment of vocabulary construction through subword tokenization algorithms and the subsequent mapping of discrete symbols into high-dimensional vector spaces via linear embeddings. Because the primary architectural mechanism-self-attention-is permutation invariant, we also detail the mathematical necessity of positional encodings to inject order into the sequence representation.

Central to our analysis is the Transformer architecture. We dissect the mechanics of self-attention, deriving the scaled dot-product formulation from first principles. This section elucidates how the model projects input vectors into Query, Key, and Value subspaces to route information based on content affinity rather than fixed topology. We further explore the structural components that enable deep signal propagation, including the interactions between position-wise feed-forward networks, layer normalization, and residual connections. Special attention is given to the implementation of causal masking, a lower-triangular constraint on the attention matrix that ensures the model remains strictly autoregressive during parallelized training.

Following the architectural decomposition, the discussion shifts to the probabilistic interpretation of the model's output. We analyze the projection of latent states onto the vocabulary and the application of the softmax function to yield a normalized probability simplex. The training process is framed as a Maximum Likelihood Estimation problem, where we derive the cross-entropy loss function used to minimize the divergence between the model's predicted distribution and the empirical distribution of the training corpus. We also address the numerical stability of these operations and the interpretation of perplexity as an exponentiated information-theoretic metric.

Finally, we address the mathematics of inference. The generated probability distributions are utilized in decoding algorithms that determine the final textual output. We contrast deterministic strategies, such as greedy decoding, with stochastic methods, including top-k and nucleus (top-p) sampling. This includes an analysis of temperature scaling and its role in modulating the entropy of the next-token distribution.

By synthesizing these components-from the initial tokenization to the final iterative decoding loop-this volume offers a comprehensive technical derivation of the logic driving Large Language Models. The aim is to provide a self-contained mathematical reference that connects the theoretical properties of probabilistic modeling with the concrete matrix operations of the Transformer architecture.

Chapter 1

Foundations of Probabilistic Text Modeling

Before "Transformers" are architectures, they are probability models: distributions over strings represented as token sequences. This chapter builds the exact measure-theoretic-and-combinatorial scaffolding needed to talk precisely about "next-token prediction" as a well-defined conditional distribution on a finite discrete support. By the end, you will be able to write down what an LLM is-independent of any neural network implementation-and see how tokenization quietly determines the sample space on which the entire theory lives.

1.1. Autoregressive Language Models as Chain-Rule Factorizations on Discrete Sequence Spaces

The fundamental object of study in probabilistic text modeling is a probability distribution over sequences of discrete symbols. To treat language mathematically, we first fix the sample space. Let V be a finite set of atomic symbols, which we call the *vocabulary*. The set of all finite sequences formed by concatenating elements of V is denoted by V^*. A language model is, in the most rigorous sense, a probability measure P defined over V^* such that the probabilities of all valid sequences sum to one.

While V^* is a countably infinite set, the high dimensionality of sequences makes direct modeling of the joint distribution $P(x)$ intractable. We cannot simply construct a histogram of all possible sentences. Instead, we rely on the chain rule of probability to decompose the joint distribution into a product of conditional distributions. For any sequence $x = (x_1, x_2, \dots, x_T)$ of length T, the joint probability is exactly:

$$P(x_{1:T}) = P(x_1) \cdot P(x_2 \mid x_1) \cdot P(x_3 \mid x_1, x_2) \cdots P(x_T \mid x_{1:T-1})$$

Using the notation $x_{<t}$ to represent the history $(x_1, \dots, x_{t-1})$ and defining $x_{<1}$ as an empty context, this factorization is written compactly as:

$$P(x_{1:T}) = \prod_{t=1}^{T} P(x_t \mid x_{<t})$$

This decomposition leads to the definition of an *autoregressive* language model. An autoregressive model does not necessarily imply a specific causal mechanism in the data generation process; rather, it is a modeling strategy that reduces the problem of estimating a joint distribution over high-dimensional objects (sequences) to the problem of estimating a series of conditional distributions over a discrete set V.

It is a common misconception that the autoregressive factorization imposes a restrictive assumption on the dependencies between tokens. The chain rule is an identity that holds for any joint distribution, provided the conditional probabilities are well-defined. Factoring a distribution left-to-right (predicting x_t from past $x_{<t}$) is mathematically equivalent to factoring it right-to-left (predicting x_t from future $x_{>t}$) or in any arbitrary permutation, in the sense that the resulting product equals the joint likelihood. The choice of the left-to-right order is pragmatic: it aligns with the temporal production of speech and text, enabling online generation where the model produces output token-by-token as it consumes it.

The transition from a theoretical decomposition to a practical model involves *parameterization*. We define a parametric function f_θ, such as a neural network, which maps a context $x_{<t}$ to a probability distribution over the vocabulary V. We denote the model's approximation of the true conditional distribution as:

$$p_\theta(x_t \mid x_{<t}) = f_\theta(x_{<t})_{x_t}$$

Here, $f_\theta(x_{<t})$ outputs a vector in the simplex $\Delta^{|V|-1}$, representing the categorical distribution over the next token. The joint probability assigned by the model to a sequence is then the product of these parameterized conditionals.

Crucially, the assumption made by modern Large Language Models (LLMs) is not the factorization itself, but the parameter sharing inherent in θ. We use the same function f_θ at every time step t. This assumes a form of stationarity or translation invariance in the conditional logic: the rules governing how a verb follows a noun phrase are presumed to be consistent regardless of whether the noun phrase appears at the beginning of the sequence or after a thousand other tokens. While the context $x_{<t}$ changes, the mechanism f_θ processing that context remains constant.

Variable Lengths and Termination

A rigorous probabilistic definition must account for the fact that sequences in natural language have variable lengths. The set V^T represents sequences of exactly length T, but V^* is the union $\bigcup_{T=0}^{\infty} V^T$. For p_θ to be a valid probability measure on V^*, the sum of probabilities over all possible sequences must equal 1.

The standard solution in autoregressive modeling is to augment the vocabulary with a special termination token, denoted as <eos> ("End of Sequence"). The generation process at each step t includes the possibility of selecting $x_t =$ <eos>. If <eos> is selected, the sequence terminates.

Formally, we consider the sample space to be the set of all sequences that end with <eos> and contain no internal <eos> tokens. Let such a sequence be $x = (x_1, \dots, x_{T-1}, \texttt{<eos>})$. Its probability is:

$$p_\theta(x) = \left(\prod_{t=1}^{T-1} p_\theta(x_t \mid x_{<t}) \right) \cdot p_\theta(\texttt{<eos>} \mid x_{<T})$$

Under this formulation, the "length" of the sequence T is not an external parameter but a random variable implicitly defined by the first index t for which $x_t =$ <eos>. For the distribution to be valid (i.e., non-defective), the model must place 0 probability on the event that the sequence never terminates. In practice, constraints on maximum sequence length or slight biases in training data ensure that the probability mass "leaks" into the <eos> state eventually, ensuring $\sum_{x \in \mathcal{X}_{\text{valid}}} p_\theta(x) = 1$.

Context as a Stochastic Process

The autoregressive framework treats the context $x_{<t}$ as a random variable. At step t, the history $X_{<t}$ is the realization of the random vari-

ables $X_1, \ldots, X_{t-1}$ generated by the previous steps of the model. Consequently, the conditional distribution $p_\theta(\cdot \mid X_{<t})$ is a stochastic kernel.

This perspective is essential for understanding inference. When we sample from an LLM, we are executing a stochastic process. The uncertainty at step t compounds over time. If the model assigns probability 0.9 to "The cat sat on the" followed by "mat", and 0.1 to "hat", the context for step $t+1$ will be different depending on this roll of the die. This branching nature creates a tree of possible futures, where the probability of any specific path is the product of the edge probabilities.

Decoding algorithms, such as Beam Search or Nucleus Sampling, are essentially search strategies or sampling techniques operating on this probabilistically generated tree. They attempt to find sequences x with high joint probability $p_\theta(x)$ or to sample x such that the distribution of samples approximates $p_\theta(x)$.

Conditional Language Modeling

The framework extends naturally to conditional language modeling, where we generate a sequence x given some input context or "prompt" y. In the autoregressive view, this is handled by simply initializing the history with y. If $y = (y_1, \ldots, y_M)$ and we wish to generate $x = (x_1, \ldots, x_N)$, the joint probability of the completion x given y is:

$$p_\theta(x \mid y) = \prod_{t=1}^{N} p_\theta(x_t \mid y, x_{<t})$$

Here, the prefix y is treated as fixed evidence. The model f_θ processes the concatenated sequence. In terms of implementation, the internal state (such as Key-Value caches in Transformers) corresponding to y is computed once, and the autoregressive generation proceeds for $x_1, x_2, \ldots$ by extending the context.

This unification of unconditional and conditional modeling is a hallmark of the LLM paradigm. By treating the prompt y simply as the initial segment of a longer sequence $y \circ x$, the same chain-rule factorization applies. The model learns to predict the next token regardless of whether the preceding tokens were provided by a user (the prompt) or generated by the model itself. This property relies on the assumption that the conditional distribution $p(x_t \mid \dots)$ behaves consistently across both regimes, an assumption justified by training on massive corpora where arbitrary substrings serve as contexts for subsequent tokens.

We have thus established the autoregressive language model as a mechanism that defines a normalized probability distribution over the space of terminated discrete sequences. By factorizing the joint likelihood into a product of conditionals, we transform the impossible task of learning a distribution over infinite strings into the tractable task of learning a next-token classifier.

1.2. Vocabularies, Token IDs, and the Geometry of Discrete Support

Probabilistic language modeling is distinguished from other signal processing tasks—such as audio or image generation—by the strictly discrete nature of its domain. While an image is naturally represented as a grid of continuous intensity values and audio as a sequence of continuous amplitudes, text is composed of categorical symbols. The fundamental atomic unit of a language model is not the character or the byte, but the *token*. Before any neural architecture or probability rule can be defined, we must rigorously specify the sample space: the universe of all possible outputs the model can generate. This specification begins with the construction of a finite vocabulary and the mapping of natural language strings onto sequences of integers.

Let Σ denote the set of base characters available in the source text (e.g., the set of all valid Unicode code points). While Σ is finite, the set of all possible strings Σ^* is countably infinite. A language model does not typically operate directly on Σ^*. Instead, we define a vocabulary V, a finite set of distinct symbol types. The size of this set, $|V|$, is a critical hyperparameter, typically ranging from 30,000 to over 100,000 in modern large language models. The elements of V are the tokens. These may correspond to single characters, whole words, or subword units, but to the model, they are merely abstract distinct entities.

To mathematically formalize the interface between raw text and the model, we assume the existence of a deterministic tokenization function $\tau : \Sigma^* \to V^*$. This function maps a string of characters s to a sequence of tokens $x = (x_1, x_2, \ldots, x_T)$, where each $x_t \in V$. Conversely, there exists a detokenization function $\tau^{-1} : V^* \to \Sigma^*$ that reconstructs the string. While τ defines how text is segmented (a topic covered in detail in the context of compression algorithms), for the purpose of defining the probability space, we treat τ as a fixed bijection between the semantic space of text strings and the combinatorial space of token sequences.

Operationally, neural networks do not process symbolic elements directly. We therefore establish a canonical bijection between the vocabulary V and the set of integers $\mathcal{I} = \{0, 1, \ldots, |V| - 1\}$. We refer to these integers as *Token IDs*. This bijection allows us to define the model's inputs and outputs numerically. If we order the vocabulary as $V = \{v_0, v_1, \ldots, v_{|V|-1}\}$, then a sequence of tokens $x_{1:T}$ is isomorphic to a sequence of integers $i_{1:T}$ where i_t is the index of x_t.

The transition from symbols to integers is not merely an implementation detail; it defines the geometry of the input and output spaces. The input to the model at time step t, x_t, is technically a categorical variable. In linear algebraic terms, we represent this variable as a one-hot vector $\mathbf{e}_{x_t} \in \{0, 1\}^{|V|}$, where the component corresponding to the token ID is

1 and all others are 0. The input space for a single token is thus the set of standard basis vectors in $\mathbb{R}^{|V|}$. This discrete, sparse representation is immediately projected into a continuous, dense vector space via an embedding matrix $E \in \mathbb{R}^{|V| \times d}$, effectively replacing the integer ID with a learned vector representation. However, the *output* of the model remains firmly rooted in the discrete geometry of the vocabulary.

The prediction of the next token x_{t+1} given a context $x_{1:t}$ is a classification problem with $|V|$ classes. The model output is a conditional probability distribution $p(\cdot \mid x_{1:t})$, which is a point in the $(|V|-1)$-simplex, denoted $\Delta^{|V|-1}$. The simplex is defined as:

$$\Delta^{|V|-1} = \left\{ \mathbf{p} \in \mathbb{R}^{|V|} \;\middle|\; p_k \geq 0, \sum_{k=0}^{|V|-1} p_k = 1 \right\}.$$

Geometrically, the language model is a function that maps a discrete path in sequence space (the context) to a point on this continuous simplex. The vertices of the simplex correspond to deterministic distributions where the mass is concentrated entirely on a single token. The center of the simplex corresponds to the uniform distribution, representing maximum entropy or total uncertainty.

This geometric perspective clarifies the role of the "logits" often discussed in implementation. A neural network typically produces an unnormalized vector $\mathbf{z} \in \mathbb{R}^{|V|}$ (the logits) rather than directly producing probabilities. The mapping from the unbounded space of logits to the simplex is performed by the softmax function $\sigma : \mathbb{R}^{|V|} \to \Delta^{|V|-1}$, defined component-wise:

$$\sigma(\mathbf{z})_k = \frac{\exp(z_k)}{\sum_{j=0}^{|V|-1} \exp(z_j)}.$$

The use of logits is preferred numerically and theoretically because the simplex is a bounded manifold, whereas the parameters of a neural network naturally reside in unbounded Euclidean space. The softmax function provides a smooth, differentiable bijection between $\mathbb{R}^{|V|}$

(modulo translation by a constant vector) and the interior of the simplex.

A rigorous definition of the sample space must also account for control codes, or special tokens. The vocabulary V is partitioned into semantic tokens (representing text content) and special tokens (representing structural information). The most critical of these are the Beginning-of-Sequence (<bos>) and End-of-Sequence (<eos>) tokens.

The <bos> token is essential for defining the initial distribution $p(x_1)$. In an autoregressive framework, every prediction is conditional. To predict the first actual token of a sequence, we condition on the start symbol: $p(x_1 \mid \texttt{<bos>})$. This allows the unified chain rule formulation $p(x_t \mid x_{<t})$ to apply even at $t = 1$.

The <eos> token is even more fundamental: it transforms the space of fixed-length sequences into a space of variable-length sequences that form a valid probability space. Without an explicit termination symbol, a language model defined by next-token probabilities would describe a stochastic process on infinite sequences, V^∞. While mathematically valid, this does not match natural language, which consists of finite discrete utterances. To model finite sequences, we augment the vocabulary V to include <eos> and define the sample space as the set of all sequences that end with <eos> and contain no other <eos> symbols. Let this set be $V^\dagger$.

$$V^\dagger = \{(x_1, \ldots, x_T) \in V^T \mid T \geq 1,\ x_T = \texttt{<eos>},\ \forall t < T,\ x_t \neq \texttt{<eos>}\}.$$

Under this definition, the "stop generating" decision is just another branch in the categorical distribution. At any step t, the model assigns a probability mass $p(\texttt{<eos>} \mid x_{<t})$ to the event that the sequence terminates. If the sum of these termination probabilities over all possible paths accumulates to 1, the model defines a valid probability mass function over finite strings. If the model assigns zero probability to <eos> everywhere, it degenerates into a process that produces infinite babble.

The inclusion of these special tokens means that the token ID space is heterogeneous. If we map V to integers $0 \ldots |V| - 1$, we might reserve ID 0 for <pad>, ID 1 for <bos>, and ID 2 for <eos>. The <pad> token deserves specific attention. Unlike <bos> and <eos>, which are part of the probabilistic event space, <pad> is a computational artifact. It is used to align sequences of different lengths into a rectangular tensor for batch processing.

Consider a batch of B sequences with lengths $L_1, L_2, \ldots, L_B$. To process this batch efficiently on hardware accelerators (GPUs/TPUs), we construct a tensor of shape (B, L_{max}), where $L_{\text{max}} = \max_i L_i$. Indices beyond the valid length L_i for the i-th sequence are filled with the <pad> ID. Crucially, the model must be prevented from attending to or learning from these padding tokens. This is achieved via masking mechanisms in the attention layers and loss functions. The probability distribution is effectively undefined (or irrelevant) at padding positions. In the loss calculation, we multiply the log-likelihood of each token by a binary mask $m_t \in \{0, 1\}$, where $m_t = 0$ if $x_t =$ <pad>. This ensures that the model's geometry is not distorted by the arbitrary requirement of rectangular memory layout.

The integer representation of tokens also dictates the dimensions of the model's final layer. The final linear projection maps the hidden state $h_t \in \mathbb{R}^d$ to the logits $z_t \in \mathbb{R}^{|V|}$. This matrix, often denoted W_{head}, has shape $d \times |V|$. For a large vocabulary (e.g., $|V| = 100{,}000$) and a moderate hidden dimension ($d = 4096$), this single matrix contains over 400 million parameters. This highlights a tension in language modeling: increasing the vocabulary size $|V|$ makes the sequence length T shorter (as each token carries more information), but it linearly increases the computational cost of the final projection and the softmax normalization.

We can visualize the interplay between text, token IDs, and special tokens with a diagram that traces the transformation of a raw string into the sequence space $V^{\dagger}$.

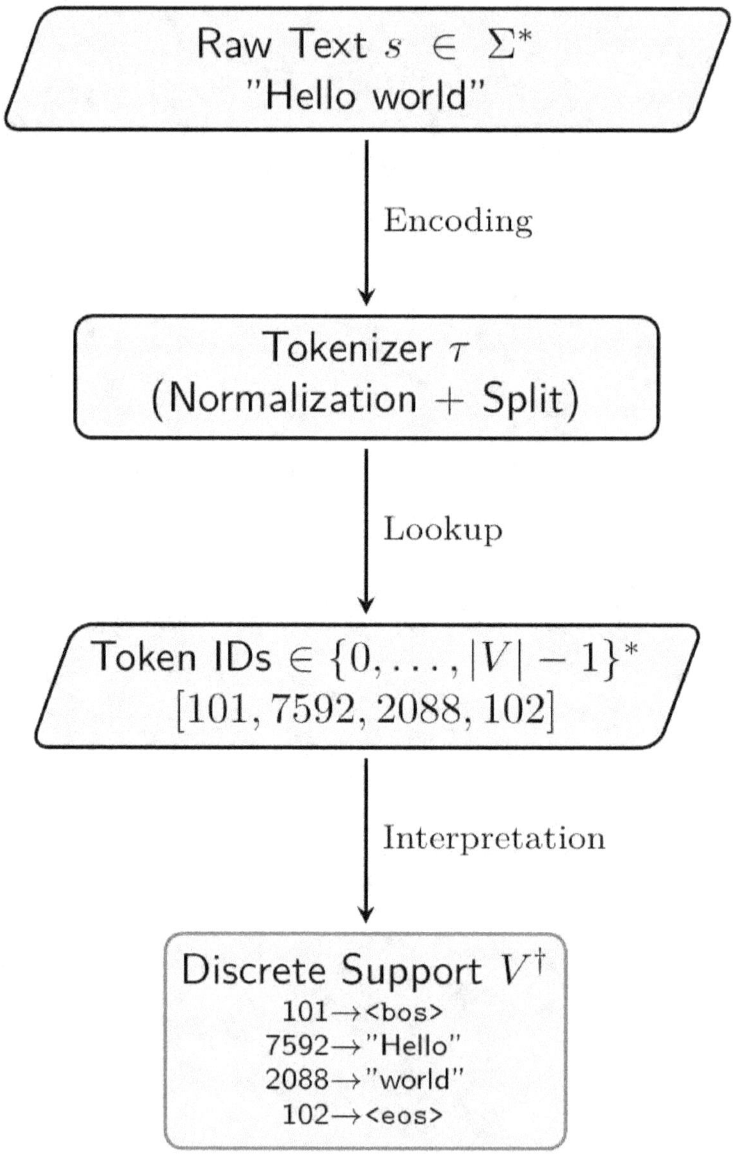
Raw Text $s \in \Sigma^*$
"Hello world"
Encoding
Tokenizer τ
(Normalization + Split)
Lookup
Token IDs $\in \{0, \ldots, |V| - 1\}^*$
$[101, 7592, 2088, 102]$
Interpretation
Discrete Support $V^\dagger$
101→<bos>
7592→"Hello"
2088→"world"
102→<eos>

The diagram illustrates the pipeline. The raw string "Hello world" is first normalized (e.g., lowercased or Unicode-normalized) and then segmented. The tokenizer inserts special tokens: a <bos> at the start (ID 101) and an <eos> at the end (ID 102). The resulting list of integers is the actual object the model manipulates. The "Discrete Support" node emphasizes that the validity of a sequence is determined by this specific structure—starting with start-of-sentence, consisting of valid vocabulary indices, and terminating with end-of-sentence.

It is important to understand the magnitude of the combinatorial space V^T. Even for a modest vocabulary of $|V| = 50{,}000$, the number of possible sequences of length $T = 20$ is $50{,}000^{20} \approx 10^{94}$. This number exceeds the number of atoms in the observable universe by dozens of orders of magnitude. The "geometry" of this support is a sparse, high-dimensional lattice. Most points in this space correspond to nonsensical noise (e.g., "the the the table table"). The task of the language model is to identify the vanishingly thin manifold within this hypergrid that corresponds to linguistically valid and semantically coherent sequences, and to assign it high probability mass.

This sparsity has profound implications for how we interpret the model's output. When we sample from $p(x_{t+1} \mid x_{1:t})$, we are sampling from a categorical distribution over 50,000 mutually exclusive events. If the model is uncertain, the probability mass may be spread thin over thousands of plausible candidates (a "flat" distribution on the simplex). If the model is confident, the mass concentrates on a few vertices (a "peaked" distribution). The entropy of the next-token distribution,

$$H(x_{t+1} \mid x_{1:t}) = -\sum_{v \in V} p(v \mid x_{1:t}) \log p(v \mid x_{1:t}),$$

measures the local branching factor of the sequence. A low entropy implies the next token is determined or highly constrained (e.g., completing a common idiom), while high entropy implies a junction point in the narrative where many continuations are possible.

Finally, we must address the concept of "Out-of-Vocabulary" (OOV) tokens. In older NLP systems, vocabulary V was often a fixed list of common words, and any word not in the list was mapped to a special `<unk>` (unknown) token. This introduced a loss of information; once a word became `<unk>`, its identity was irretrievably lost, and the model could not generate it. Modern subword tokenization schemes (discussed in the subsequent section) largely eliminate `<unk>` by ensuring that the base vocabulary includes all individual characters or bytes. Thus, any string can be decomposed into smaller units in V. In this paradigm, V forms a *universal* covering of the semantic space Σ^*, ensuring that the support of the probability model effectively covers all representable text.

By fixing V, mapping text to integer sequences $V^\dagger$, and treating the outputs as categorical distributions on the simplex, we define the formal arena for language modeling. The "geometry" is not continuous in the sense of Euclidean distance between input tokens—ID 105 is not necessarily "closer" to ID 106 than to ID 5000—but rather defined by the learned probability masses on the simplex vertices. The autoregressive factorization is the mechanism that allows us to traverse this combinatorial space one step at a time, converting a joint distribution over the unimaginable magnitude of V^T into a manageable sequence of $|V|$-way classification problems.

1.3. Subword Tokenization as Learned Discrete Compression: BPE/WordPiece, Likelihood Surrogates, and Entropy Trade-offs

The definition of a language model as a distribution over sequences $x \in V^*$ relies entirely on the prior existence of a finite vocabulary V.

In the previous analysis of autoregressive factorizations, V was treated as a fixed, axiomatic set of symbols. However, in modern probabilistic text modeling, V is not given; it is learned. The construction of V and the associated encoding mapping enc : $\Sigma^* \to V^*$ constitute the first stage of modeling, commonly referred to as subword tokenization. This process is fundamentally a data-driven compression scheme designed to balance the expressivity of the discrete sample space against the computational cost of sequence processing.

Tokenization transforms the raw data-sequences of Unicode characters or bytes-into the discrete event space where the probabilistic chain rule applies. If the mapping is bijective (reversible), the probability assigned to a token sequence $p_\theta(x_{1:T})$ can be rigorously translated back to a density over the original text. If the mapping is lossy (e.g., due to aggressive normalization or ignoring whitespace), the model defines a distribution over a simplified semantic space, and the connection to the raw signal is severed. Therefore, subword algorithms are not merely preprocessing steps but structural priors that determine the granularity of the data, the sparsity of the statistics, and the effective length of dependencies.

The Open-Vocabulary Requirement and Byte-Level Fallback

A robust probabilistic model must assign non-zero probability to any valid input string. Traditional word-level vocabularies failed this requirement; any word not in V was mapped to a special unknown token `<unk>`, collapsing the massive tail of rare words, misspellings, and novel compounds into a single singularity in the probability space. This creates a "blind spot" where the model cannot distinguish between completely different rare events.

Subword modeling resolves this by enforcing the *open-vocabulary property*: the base alphabet of the tokenizer must be sufficient to generate any string in Σ^*. While early approaches used character sets, the dominance of Unicode (with over 149,000 defined characters) makes a full character-level vocabulary unwieldy. Instead, modern systems invariably retreat to the byte level. By treating text as a stream of UTF-8 bytes, the base vocabulary size is fixed at 256. Any string, regardless of language or encoding, is decomposable into bytes.

However, modeling raw bytes is computationally inefficient. The entropy rate of text per byte is low, meaning each individual byte carries little semantic information, and the sequence length T for a given document becomes very large. Since the computational cost of Transformer-based architectures scales quadratically (or at best linearly) with T, pure byte-level modeling wastes capacity on local syntactic composition (assembling "t-h-e") rather than semantic reasoning. Subword algorithms bridge this gap by learning a vocabulary V that includes the base bytes plus thousands of frequent variable-length byte sequences (subwords), effectively compressing the stream.

Byte Pair Encoding (BPE) as Greedy Compression Byte Pair Encoding (BPE) is the most prevalent algorithm for constructing such vocabularies. Originally a compression technique, BPE constructs V through an iterative, greedy process based on co-occurrence frequencies.

Let the corpus be represented initially as a sequence of base symbols (e.g., bytes or characters). We define the vocabulary V_0 as the set of all unique base symbols. In each iteration k, the algorithm counts all adjacent pairs (u, v) in the current segmented corpus. The most frequent pair is designated as a new symbol $w_{new} = uv$. We update the vocabulary $V_k = V_{k-1} \cup \{w_{new}\}$ and replace all occurrences of the pair (u, v) with w_{new}. This process repeats until $|V|$ reaches a target capacity (typically between 32,000 and 100,000).

The BPE algorithm can be implemented efficiently without re-scanning the full corpus at every step. By maintaining a linked list of symbol occurrences and a priority queue of pair counts, the updates remain local. The following Python snippet illustrates the core merge logic on a simplified counting dictionary:

```
def get_stats(vocab_counts):
    """Compute frequencies of adjacent pairs."""
    pairs = {}
    for word, freq in vocab_counts.items():
        symbols = word.split()
        for i in range(len(symbols)-1):
            pair = (symbols[i], symbols[i+1])
            pairs[pair] = pairs.get(pair, 0) + freq
    return pairs

def merge_vocab(pair, vocab_counts):
    """Replace pair (A, B) with new symbol AB in all words."""
    v_out = {}
    bigram = ' '.join(pair)
    replacement = ''.join(pair)
    import re
    p = re.compile(r'(?<!\S)' + re.escape(bigram) + r'(?!\S)')
    for word, freq in vocab_counts.items():
        # Apply merge: 'x y z' -> 'x yz' if pair is ('y','z')
        w_new = p.sub(replacement, word)
        v_out[w_new] = freq
    return v_out
```

The result of BPE is a set of merge rules. At inference time, these rules are applied deterministically to encode new text. The greedy nature of BPE implies that it does not optimize a global objective function over the corpus; rather, it optimizes the immediate reduction in sequence length by merging the most common adjacent symbols. This aligns with a compression objective: replacing frequent pairs with single symbols minimizes the total number of tokens required to represent the data.

Crucially, standard BPE implementations do not cross word boundaries (defined by whitespace). This constraint simplifies the tokenizer but requires a pre-tokenization step to split text into words (or "pre-tokens"). Recent approaches, such as those used in GPT-2 and GPT-3, treat spaces as just another character (or byte), allowing BPE to merge across boundaries (e.g., capturing the frequent sequence " in the"). This effectively absorbs the "word segmentation" problem into the general compression problem, rendering the mapping fully reversible without specialized detokenization logic.

Likelihood-Based Segmentation: WordPiece and Unigram LM While BPE relies on frequency, other schemes optimize a probabilistic objective directly. WordPiece, used in BERT, employs a similar merge strategy to BPE but selects the pair to merge based on likelihood rather than raw count.

Given a current vocabulary and a unigram language model (where each token t occurs with probability $p(t)$), the likelihood of the corpus $C = \{t_1, \ldots, t_N\}$ is $\prod_{i=1}^{N} p(t_i)$. Merging a pair (u, v) into z changes the likelihood. The local change depends on the count of the pair relative to the counts of the individual parts. Specifically, WordPiece chooses to merge the pair that maximizes the score $\frac{p(uv)}{p(u)p(v)}$, which is equivalent to maximizing the mutual information between the adjacent symbols. This tends to prioritize merging symbols that are strongly correlated, even if their absolute frequency is lower than other pairs.

A more rigorous probabilistic framework is provided by the Unigram Language Model tokenizer (often implemented in SentencePiece). Unlike BPE and WordPiece, which start small and build up, the Unigram approach starts with a massive heuristic vocabulary and prunes it down. It assumes that for any text sequence X, there exists a set of valid segmentations $S(X)$. A simple unigram model assigns a probability to a segmentation $s = (x_1, \ldots, x_M)$ as $P(s) = \prod_{i=1}^{M} p(x_i)$. The objective is to find the optimal vocabulary that maximizes the marginal likelihood of the data $\sum_{s \in S(X)} P(s)$ (or, in the Viterbi approximation, the probability of the most likely segmentation).

During training, the Unigram tokenizer alternates between optimizing the unigram probabilities $p(x_i)$ (via Expectation-Maximization) and pruning symbols that contribute least to the overall likelihood. This approach has a distinct theoretical advantage: it explicitly acknowledges that segmentation is ambiguous. While BPE defines a single deterministic segmentation path, the Unigram model allows for probabilistic sampling of segmentations. This enables *subword regularization*, where the model is trained on multiple valid segmentations of the same text (e.g., "probabilistic" vs. "prob" + "abilistic"), improving robustness to noise and vocabulary mismatch.

Entropy Trade-offs and the Granularity Spectrum The choice of vocabulary size $|V|$ dictates the position of the model on a granularity spectrum, governing the trade-off between the entropy of the prediction distribution and the effective length of the sequence.

Let $H(X)$ be the entropy of the underlying text distribution. If we view the text as a stochastic process, the information content is invariant to the encoding scheme (by the data processing inequality, we cannot create information, only preserve it). However, the *distribution of entropy across steps* changes drastically.

- *Character/Byte Level (Small $|V|$, Large T):* The sequence is long. The entropy per step is low (at most $\log_2 256 = 8$ bits, but empirically much lower, often < 1.5 bits/byte for English). The conditional distribution $p(x_t \mid x_{<t})$ is sharp; the model only needs to predict the next letter, which is often syntactically constrained. However, semantic information is diffused across many steps. The model must propagate state over hundreds of steps to resolve simple dependencies (e.g., subject—verb agreement).
- *Word/Phrase Level (Large $|V|$, Small T):* The sequence is short.

The entropy per step is high. Predicting the next "word" involves choosing from a vast set of semantic possibilities. The distribution $p(x_t \mid x_{<t})$ is flatter and heavy-tailed. The computational burden shifts from maintaining long-term memory to computing massive softmax normalizations and handling data sparsity (many tokens in V will appear rarely).

Subword tokenization aims for the "sweet spot" where T is short enough to fit context windows (typically 2048 to 32k tokens) while $|V|$ is small enough to ensure dense embeddings and efficient softmax computation.

We can quantify this trade-off using the concept of average fertility or compression rate. If a raw string has length L in bytes and is tokenized into T tokens, the compression ratio is L/T. BPE typically achieves a ratio of 3–4 bytes per token for English text. This ratio is not uniform; frequent words are 1 token, while random strings revert to 1 byte per token. This variable-length encoding acts as an impedance matcher: it allocates unique symbols to high-information, frequent semantic units, and relies on composition for low-frequency or high-entropy sequences.

Reversibility, Normalization, and the Lossless Ideal The mathematical contract of the language model $p_\theta(x)$ is that it models the distribution of tokens. To interpret this as a distribution over text, the tokenization must be reversible. Formally, if $\mathcal{D} : V^* \to \Sigma^*$ is the decoding function, we require that for any valid string $s \in \Sigma^*$, $\mathcal{D}(\text{enc}(s)) \approx s$.

Strict equality is often compromised by normalization. Before tokenization, text is usually normalized (e.g., Unicode NFKC, lowercasing, stripping accents). If the normalization is non-injective (mapping multiple distinct raw strings to the same normalized form), the model loses the ability to distinguish them, and the distribution p_θ is defined effectively on the quotient space of strings under the normalization equivalence relation.

For modern general-purpose models (e.g., code generation, multilingual tasks), lossless reconstruction is critical. This is achieved by:

- Preserving Whitespace: Unlike standard English prose tokenization which might discard extra spaces, code and formatted text require exact whitespace preservation. SentencePiece and GPT-2/3 tokenizers handle this by treating space as a literal symbol (often visualized as _ or \u0120) that is part of the token.
- Byte-Level Fallback: By falling back to bytes for any sequence not in the merge table, the tokenizer ensures that no "unknown" characters exist. This guarantees that enc(s) is always defined and $\mathcal{D}$ can reconstruct s byte-for-byte.

When these conditions are met, the probability assigned to a string s is exactly the probability of its unique tokenization $x_{1:T}$:

$$p_{\text{text}}(s) = p_\theta(x_{1:T}).$$

However, if the tokenizer is probabilistic (like Unigram LM with sampling) or if multiple tokenizations map to the same string (ambiguity), the probability of the string is the marginal sum over all valid segmentations:

$$p_{\text{text}}(s) = \sum_{x \in \text{enc}^{-1}(s)} p_\theta(x).$$

Most autoregressive models implicitly assume a canonical, deterministic segmentation (the "Viterbi" path) during training, effectively optimizing a lower bound on the log-likelihood of the string.

Pathologies and Domain Mismatch The tokenization map is learned from a specific training corpus. This introduces a "vocabulary bias." A tokenizer trained on English prose will learn merges like "ing", "tion", " the". If applied to Python code, it may inefficiently segment keywords (e.g., "import" might be split if not frequent in prose) or variable names, leading to inflated sequence lengths.

This domain mismatch manifests as a degradation in the compression ratio. If the distribution of text shifts such that the learned subwords are no longer frequent, the tokenizer regresses towards character-level splitting. This increases the effective context cost for the model, as the limited window size (in tokens) covers fewer actual bytes of information. Furthermore, it alters the probability landscape: the model must now perform more steps of uncertainty reduction to express the same semantic content.

A particularly subtle pathology arises in "glitch tokens." These are tokens that exist in the vocabulary (perhaps due to noise in the tokenizer's training data) but are vastly under-represented in the model's training data. During generation, if the model stumbles upon such a token, the associated embeddings may be untrained, leading to undefined behavior or numerical instability.

In summary, subword tokenization defines the discrete coordinate system for the language model. It is a compression algorithm that transforms the sparse, high-dimensional space of raw text into a dense, lower-dimensional sequence space. The properties of this transformation-its reversibility, compression rate, and handling of rare events-are baked into the probability values $p_\theta(x)$ and cannot be undone by the model architecture. The vocabulary V is the static hard-coding of prior knowledge about the structure of the data, determining the resolution at which the model views the world.

Chapter 2

Vector Space Representations

Tokens are integers; transformers are smooth functions optimized by gradient descent. This chapter bridges that gap by treating "token identity" as a basis-vector selection problem (one-hot) that is equivalent to a linear map, then extending the representation to sequences by injecting order through positional signals. The core message is that most of an LLM's expressivity begins with a few precise shape constraints and invariances in high-dimensional vector spaces-and you can reason about them rigorously with linear algebra alone.

2.1. Token Embeddings as Basis Selection: One-Hot Vectors, Lookup Tables, and Linear Operator Equivalence

To understand the mechanics of large language models, we must first rigorously define the boundary between the discrete space of symbols and the continuous space of vector representations. A token is an element of a finite vocabulary set V with cardinality $|V| = n$. While tokens are often stored and processed as integers $i \in \{0, \ldots, n-1\}$ for efficiency, integers imply an ordinal relationship (e.g., $2 > 1$) that does not exist in the semantic space of language. "Apple" is not greater than "Aardvark" in any inherent sense, nor is "Zebra" the sum of "Apple" and "Aardvark."

The canonical algebraic representation of a categorical variable with n levels is the standard basis vector in $\mathbb{R}^n$. Let $e_i \in \mathbb{R}^n$ be a vector with a 1 in the i-th position and 0 elsewhere. This set of vectors $\{e_0, \ldots, e_{n-1}\}$ forms an orthonormal basis for the token space. The dot product $e_i^\top e_j = \delta_{ij}$ (Kronecker delta) formally encodes the distinctness of tokens: every token is orthogonal to every other token, and the distance between any two distinct tokens is constant ($\sqrt{2}$). This high-dimensional, sparse representation cleanly separates the combinatorics of token selection from the geometry of semantic meaning.

The embedding layer is a learned linear map $\mathcal{E} : \mathbb{R}^n \to \mathbb{R}^d$ that projects these high-dimensional basis vectors into a lower-dimensional, continuous vector space $\mathbb{R}^d$, where $d \ll n$. This map is parameterized by an embedding matrix $E \in \mathbb{R}^{n \times d}$. Applying the linear map to a basis vector e_i corresponds to matrix multiplication:

$$\mathbf{h}_i = e_i^\top E$$

Here, we adhere to the row-vector convention common in deep learning implementations, where data vectors are rows. Because e_i has a single non-zero entry at index i, this matrix multiplication simplifies to selecting the i-th row of E. This establishes the fundamental equivalence: *embedding lookup is matrix multiplication by a one-hot vector.*

While modern deep learning frameworks implement embeddings as efficient array lookups (indexing) to avoid storing and multiplying by distinct sparse matrices, the linear operator view is not merely a theoretical curiosity; it is the correct mathematical framework for analyzing gradients, rank, and parameter constraints. The following code snippet demonstrates this exact numerical equivalence using PyTorch, highlighting that the "lookup table" is simply a linear layer applied to sparse inputs.

```
import torch
import torch.nn.functional as F

# Configuration
vocab_size = 5
embed_dim = 4
torch.manual_seed(42)

# The Embedding Matrix E (n x d)
E = torch.randn(vocab_size, embed_dim, requires_grad=True)

# A sequence of token indices: [1, 3, 0]
indices = torch.tensor([1, 3, 0])

# Method A: Efficient Lookup (Gather)
# This is how frameworks implement it for O(1) access per token
h_lookup = F.embedding(indices, E)

# Method B: Linear Algebra (One-Hot MatMul)
# e_i^T E
one_hots = F.one_hot(indices, num_classes=vocab_size).float() # Shape
    : (3, 5)
h_matmul = one_hots @ E                                        # Shape
    : (3, 4)

# Verification of equivalence
difference = (h_lookup - h_matmul).abs().max().item()
print(f"Max absolute difference: {difference:.2e}")
```

```
# Output typically ~0.00e+00 (within float precision)
```

```
Max absolute difference: 0.00e+00
```

When we extend this to a sequence of length T, the input becomes a matrix $X \in \mathbb{R}^{T \times n}$, where each row t is the one-hot vector $e_{i_t}^\top$ corresponding to the token at position t. The embedding operation for the entire sequence is the matrix product:

$$H = XE$$

The resulting matrix $H \in \mathbb{R}^{T \times d}$ contains the dense vector representations for the sequence. This shape transformation-from $T \times n$ (sparse, high-dimensional) to $T \times d$ (dense, lower-dimensional)-is the entry point for the Transformer architecture. From this point forward, the model operates exclusively on H, treating the input as a sequence of vectors in a continuous metric space rather than a sequence of symbols.

The linear-operator perspective clarifies the mechanics of backpropagation through the embedding layer. Let L be the scalar loss function. By the chain rule, the gradient of the loss with respect to the embedding matrix E is derived from the gradient with respect to the hidden states $\nabla_H L \in \mathbb{R}^{T \times d}$. Using standard matrix calculus identities for linear maps:

$$\frac{\partial L}{\partial E} = X^\top \left(\frac{\partial L}{\partial H} \right)$$

Since X is a collection of one-hot vectors, $X^\top$ acts as a scattering operator. If the token index k appears at positions $t_1, t_2, \ldots$ in the sequence, the k-th row of the gradient matrix $\frac{\partial L}{\partial E}$ is simply the sum of the gradients $\frac{\partial L}{\partial h_t}$ for all positions t where the token k appeared:

$$\left(\frac{\partial L}{\partial E}\right)_{k,:} = \sum_{t \text{ s.t. } i_t = k} \left(\frac{\partial L}{\partial H}\right)_{t,:}$$

For any token j not present in the current batch ($j \notin \{i_1, \ldots, i_T\}$), the corresponding row in $X^\top$ is zero, and thus the gradient is zero. This *sparse update property* is a direct consequence of the orthogonality of the basis vectors. In distributed training or systems with extremely large vocabularies, this sparsity is exploited to reduce communication overhead; we only need to synchronize the rows of E that were actually accessed.

We can also analyze the gradient with respect to the input X. Mathematically,

$$\frac{\partial L}{\partial X} = \left(\frac{\partial L}{\partial H}\right) E^\top.$$

This quantity is a $T \times n$ matrix where each entry (t, k) represents the sensitivity of the loss to the presence of token k at position t. While we cannot perform gradient descent on the discrete indices directly (since X is constrained to be one-hot integers), this gradient is the basis for *adversarial token attacks* and gradient-based interpretability methods. It approximates how replacing the token at position t with token k would locally affect the loss.

The dimensionality d of the embedding space acts as a bottleneck on the rank of the learned representations. The matrix E has size $n \times d$, and typically $d \ll n$. Consequently, the maximum possible rank of E is d. This implies that the n token vectors in $\mathbb{R}^d$ must lie on a d-dimensional subspace. They cannot be linearly independent; there are linear dependencies among the token embeddings. The model must compress the semantic relationships of n concepts into a d-dimensional geometry. If d is too small (e.g., $d = 2$ for a vocabulary of 50,000), the "pigeonhole principle" of linear algebra forces dissimilar tokens to be placed near each other, collapsing distinct semantic meanings (embedding collapse). Conversely, a sufficiently large d allows the simplex of n tokens to be embedded such that semantic similarities are preserved via dot products.

This geometric view illuminates the practice of *parameter tying* (or weight sharing) between the input embedding and the output projection layer. In many Transformer implementations, the probability distribution over the next token is computed as:

$$P(y_{t+1} \mid H_t) = \text{softmax}(h_t W_{\text{out}}^\top)$$

where $W_{\text{out}} \in \mathbb{R}^{n \times d}$ projects the final hidden state back to the vocabulary size. Parameter tying imposes the constraint $W_{\text{out}} = E$. Under the linear operator view, this is not merely a heuristic to save parameters. It forces the input representation and the output decision boundary to share the same metric space. If $W_{\text{out}} = E$, the logit for token k is the dot product $h_t \cdot e_k^\top E = h_t \cdot \mathbf{h}_k^\top$. Thus, the model predicts the next token by finding the token embedding $\mathbf{h}_k$ that has the highest inner product with the current context vector h_t. This aligns the geometry of "what I am reading" with "what I am writing," ensuring that the vector space remains consistent throughout the forward pass.

Finally, we must distinguish between the mathematical properties of the embedding map and the semantic properties acquired during training. The linear equivalence $H = XE$ is exact and training-independent. However, the geometric relationships between rows of E-that "king" - "man" + "woman" $\approx$ "queen"-are entirely empirical results of the optimization process. The embedding layer itself imposes no such structure; it simply provides the linear degrees of freedom (the entries of E) required for the optimizer to arrange tokens in space such that their dot products minimize the downstream loss. The power of the embedding layer lies in its role as a differentiable interface: it converts the rigid, non-differentiable discrete equality of tokens ($i = j$ or

$i \neq j$) into a smooth, differentiable similarity measure in $\mathbb{R}^d$.

2.2. Breaking Permutation Symmetry: Positional Signals as Additive, Learned, and Functional Encodings

In the previous derivation of token embeddings, we established the representation of a sequence as a matrix $H \in \mathbb{R}^{T \times d}$. This matrix is the product of a one-hot selection matrix X and the embedding matrix E. While this construction successfully captures token identity and maps it into a continuous vector space, it is deficient in one critical aspect: it lacks order.

Consider a transformation function $f : \mathbb{R}^{T \times d} \to \mathbb{R}^{T \times d}$ representing a layer in the network. If f processes each row of H independently (as in a feed-forward layer) or computes interactions based solely on content similarity (as in raw self-attention), the function is *permutation equivariant*. Formally, for any permutation matrix $\Pi \in \{0,1\}^{T \times T}$, the function satisfies:

$$f(\Pi H) = \Pi f(H)$$

Under this constraint, a sequence "A B C" is indistinguishable from "C B A" in terms of the set of output vectors produced; the outputs are simply permuted in the same manner as the inputs. Since natural language and many sequential data types rely on order to encode meaning (e.g., subject–object distinction), the representation H must be augmented to break this symmetry.

We resolve this by treating position not as metadata managed by an external iterator, but as a signal embedded directly into the vector space. We define a positional encoding function $\phi_{\text{pos}} : \mathcal{P} \to \mathbb{R}^d$, where $\mathcal{P}$ is the set of valid positions. For a token at position t, the full representation becomes a combination of the token signal h_t and the position signal

$p_t = \phi_{\text{pos}}(t)$.

The standard mechanism for this combination in Transformer architectures is addition:

$$\tilde{h}_t = h_t + p_t$$

This design choice is non-trivial. Concatenation, defined as $[h_t; p_t] \in \mathbb{R}^{d+d_{\text{pos}}}$, would preserve the distinctness of the two signals perfectly but would increase the dimensionality of the state space, requiring larger weight matrices in all subsequent layers. By choosing addition, we impose a superposition constraint: the model must learn to perform linear separations that distinguish semantic content from positional information within the same d-dimensional space.

Learned Absolute Positional Embeddings

The most direct method to implement ϕ_{pos} parallels the construction of token embeddings. We assume a maximum sequence length T_{max} and define a vocabulary of positions $\{1, \ldots, T_{\text{max}}\}$. We allocate a learnable parameter matrix $P \in \mathbb{R}^{T_{\text{max}} \times d}$, where the t-th row corresponds to the vector p_t.

For a sequence of length T, we construct a selection matrix $S \in \{0, 1\}^{T \times T_{\text{max}}}$, where $S_{t,k} = 1$ if the t-th token in the sequence is at absolute position k, and 0 otherwise. In the standard case where the sequence starts at position 1, S is simply the first T rows of the identity matrix. The input representation $\tilde{H}$ is then:

$$\tilde{H} = XE + SP$$

Here, XE provides the semantic content and SP provides the positional bias. Because P is a free parameter matrix updated via backpropagation, the model learns the optimal vector representation for "position 1," "position 2," and so forth.

This approach effectively breaks permutation symmetry but introduces two significant limitations. First, it is strictly inductive but not extrapolative with respect to position index. If the model is trained with a maximum context length $T_{\max}$, the matrix P has no defined row for position $T_{\max} + 1$. Handling longer sequences requires architectural modification or retraining. Second, there is no inherent geometric relationship between p_t and p_{t+1} other than what is induced by the training data. The model must observe examples to learn that position t is spatially close to $t + 1$.

Sinusoidal Functional Encodings

To address the lack of geometric structure and the extrapolation limit, we can define ϕ_{pos} as a fixed, continuous function. The sinusoidal encoding scheme constructs a deterministic vector for each position t using a range of frequencies.

For a dimension index $j \in \{0, \ldots, d-1\}$, the positional vector p_t is defined coordinate-wise. The frequencies follow a geometric progression, typically ranging from a wavelength of 2π to $10000 \cdot 2\pi$. The encoding is defined as:

$$p_{t,2i} = \sin\left(\frac{t}{10000^{2i/d}}\right)$$

$$p_{t,2i+1} = \cos\left(\frac{t}{10000^{2i/d}}\right)$$

where i indexes the frequency band. Each pair of dimensions $(2i, 2i+1)$ forms a 2D subspace where position t is represented as a rotation.

```
import numpy as np

def get_sinusoidal_encoding(length, dim):
    # Shape: (length, 1)
    position = np.arange(length)[:, np.newaxis]
    # Shape: (1, dim//2)
    div_term = np.exp(
```

```
        np.arange(0, dim, 2) * -(np.log(10000.0) / dim)
    )

    pe = np.zeros((length, dim))
    pe[:, 0::2] = np.sin(position * div_term)
    pe[:, 1::2] = np.cos(position * div_term)
    return pe
```

This functional form provides several representation-theoretic advantages. First, the norm of the position vector is deterministic and bounded (specifically, $\|p_t\|^2 = d/2$), ensuring that the positional signal does not overwhelm the token embedding magnitude, provided d is sufficiently large and token embeddings are properly normalized.

Second, this encoding allows linear operators to attend to relative positions. For any fixed offset k, the vector p_{t+k} at a specific frequency ω_i can be expressed as a linear transformation of p_t. Using the angle addition formulas:

$$\sin(\omega_i(t+k)) = \sin(\omega_i t)\cos(\omega_i k) + \cos(\omega_i t)\sin(\omega_i k)$$
$$\cos(\omega_i(t+k)) = \cos(\omega_i t)\cos(\omega_i k) - \sin(\omega_i t)\sin(\omega_i k)$$

In matrix notation, for the 2D subspace corresponding to frequency ω_i, the transition from position t to $t+k$ is a rotation:

$$\begin{bmatrix} p_{t+k,2i} \\ p_{t+k,2i+1} \end{bmatrix} = \begin{bmatrix} \cos(\omega_i k) & \sin(\omega_i k) \\ -\sin(\omega_i k) & \cos(\omega_i k) \end{bmatrix} \begin{bmatrix} p_{t,2i} \\ p_{t,2i+1} \end{bmatrix}$$

This rotation matrix R_k depends only on the offset k, not on the absolute position t. Consequently, a learnable weight matrix W in the network can effectively inspect relative distances by interacting with these rotation properties. If the network learns to compute dot products or similarities that align with these rotations, it achieves translation invariance in its processing of the sequence.

The Additive Space and Interference

Whether learned or functional, the choice to sum $h_t + p_t$ implies that the vector space is a shared medium for two distinct types of information. We can model this as a subspace decomposition. Ideally, the token embeddings E populate a subspace $U_{\text{sem}} \subset \mathbb{R}^d$, and the positional embeddings P populate $U_{\text{pos}} \subset \mathbb{R}^d$. If these subspaces are nearly orthogonal, simple linear projections can recover either signal with minimal interference.

However, as dimensionality constraints tighten or d becomes small relative to the complexity of the data, this orthogonality cannot be guaranteed. The interference between token identity and position is not necessarily detrimental; it creates a composite representation where "the word 'dog' at position 5" is a unique point in $\mathbb{R}^d$, distinct from "the word 'dog' at position 10". This compositionality is the fundamental requirement for the subsequent attention mechanism to operate not just on *what* was said, but *where* it appeared in the stream.

The resulting matrix $\tilde{H} \in \mathbb{R}^{T \times d}$ is the definitive input to the transformer blocks. It carries the discrete token choices chosen via E, the sequential ordering injected via P (or ϕ_{pos}), and maintains the exact shape required for the affine transformations discussed in the following section.

2.3. Affine Geometry of Hidden States: High-Dimensional Linear Maps, Biases, and Pointwise Nonlinearities

The computational backbone of deep learning architectures, particularly the Transformer, is the affine transformation followed by a pointwise nonlinearity. While attention mechanisms often dominate the conceptual discussion, the vast majority of parameters and floating-point operations in a standard Large Language Model (LLM) reside in the feed-forward networks (FFNs) and projection layers. These layers are responsible for the bulk of the associative memory and feature processing. Understanding their geometry requires moving beyond the elementary view of "multiplying by weights" to a rigorous analysis of high-dimensional affine maps, the critical role of bias terms in shaping representational manifolds, and the interaction between linear operators and activation functions that grants these models their universal approximation capabilities.

At the level of a single token representation $x \in \mathbb{R}^{d_{\text{in}}}$, a linear layer applies a transformation map $f : \mathbb{R}^{d_{\text{in}}} \to \mathbb{R}^{d_{\text{out}}}$. In its most general affine form, this map is defined by a weight matrix $W \in \mathbb{R}^{d_{\text{out}} \times d_{\text{in}}}$ and a bias vector $b \in \mathbb{R}^{d_{\text{out}}}$:

$$y = Wx + b.$$

This formulation assumes x is a column vector. However, in implementation and systems discussion, representations are almost invariably stored as row vectors to align with memory layout conventions where the token sequence dimension is the primary batching axis. If we denote the single token input as a row vector $x^\top \in \mathbb{R}^{1 \times d_{\text{in}}}$, the transformation becomes:

$$y^\top = x^\top W^\top + b^\top.$$

While mathematically equivalent, the distinction is significant for shape discipline. In the column-vector convention, W transforms the basis of the domain $\mathbb{R}^{d_{\text{in}}}$ to the codomain $\mathbb{R}^{d_{\text{out}}}$. The geometric action of W can be decomposed via Singular Value Decomposition (SVD) into a rotation, a scaling along principal axes (singular values), and a second rotation. The bias b then applies a translation. This translation does not depend on the input x; it is a constant shift of the output space. Geometrically, a linear map without bias ($y = Wx$) fixes the origin: the zero vector 0 must map to 0. An affine map ($y = Wx + b$)

shifts the origin to b. As we will see, this freedom to move the origin is essential for the expressivity of piecewise-linear networks.

When processing a sequence of tokens T, we organize the inputs into a matrix $H \in \mathbb{R}^{T \times d_{\text{in}}}$, where each row corresponds to a token embedding at a specific position. The application of the affine layer across the sequence must preserve the independence of the token positions (excluding the mixing performed by attention). This is achieved by broadcasting the same weights W and bias b across the row dimension of H. The batched affine transformation is:

$$Y = HW^{\top} + \mathbf{1}b^{\top},$$

where $Y \in \mathbb{R}^{T \times d_{\text{out}}}$ and $\mathbf{1} \in \mathbb{R}^{T}$ is the column vector of all ones. The term $\mathbf{1}b^{\top}$ is a rank-1 matrix formed by repeating the row vector $b^{\top}$ T times.

This broadcasting operation highlights a crucial symmetry: the layer is permutation equivariant with respect to the sequence dimension. If we permute the rows of H, the rows of Y are permuted identically. This confirms that the dense layers in a Transformer operate "position-wise." They possess no intrinsic notion of order or proximity; they apply the exact same function to the representation at position $t = 1$ as they do to $t = 100$. Any sensitivity to sequence order must be injected via the positional encodings discussed previously or derived from the contextual mixing in attention layers.

From a systems perspective, the computation $Y = HW^{\top}$ is a General Matrix Multiply (GEMM), which is the fundamental kernel optimized by hardware accelerators. The performance of this operation is governed by the arithmetic intensity—the ratio of floating-point operations to memory accesses. Since W is loaded once and reused for all T rows of H, a larger sequence length T increases arithmetic intensity, allowing the utilization of the GPU's compute cores rather than being bottlenecked by memory bandwidth. This provides a hardware-level justification for processing tokens in batches.

The following Python code snippet using PyTorch demonstrates the precise broadcasting semantics and the shape mechanics of the linear layer. Note that high-level frameworks often hide the explicit expansion of the bias, but the underlying linear algebra engine performs exactly the rank-1 update described.

```
import torch

# Dimensions
T = 5        # Sequence length
d_in = 4     # Input dimension
d_out = 3    # Output dimension

# Input sequence H (T, d_in)
H = torch.randn(T, d_in)

# Weight W (d_out, d_in) - typically stored this way
W = torch.randn(d_out, d_in)

# Bias b (d_out)
b = torch.randn(d_out)

# 1. Standard Linear Layer application
linear_layer = torch.nn.Linear(d_in, d_out)
linear_layer.weight.data = W
linear_layer.bias.data = b
Y_layer = linear_layer(H)

# 2. Explicit Algebraic Formulation
# Transpose W to align input dimension: (T, d_in) @ (d_in, d_out) ->
    (T, d_out)
# Bias b is (d_out), implicitly broadcast to (T, d_out)
Y_explicit = H @ W.T + b

# Verification of equality
diff = (Y_layer - Y_explicit).abs().max()
print(f"Max difference: {diff:.2e}")

# Explicit Rank-1 Bias Expansion
ones = torch.ones(T, 1)
bias_matrix = ones @ b.unsqueeze(0) # (T, 1) @ (1, d_out) -> (T,
    d_out)
Y_fully_expanded = H @ W.T + bias_matrix
```

```
Max difference: 0.00e+00
```

While the matrix multiplication $HW^\top$ provides the mixing of dimensions, the bias term b plays a specific and often underestimated geometric role. In high-dimensional spaces, the "origin" is merely a coordinate artifact. However, linear maps ($y = Wx$) are structurally tethered to this origin. The set of inputs x such that $w_i \cdot x = 0$ defines a hyperplane passing directly through the origin. Without a bias term, every neuron in a layer defines a decision boundary that must intersect at 0.

This restriction becomes problematic when we introduce nonlinearities. Consider a ReLU activation $\sigma(z) = \max(0, z)$. A neuron computes $y_i = \max(0, w_i \cdot x + b_i)$. The active region of this neuron is the half-space $w_i \cdot x + b_i > 0$. The boundary of this region is the hyperplane $w_i \cdot x + b_i = 0$. The distance of this hyperplane from the origin is determined by $b_i / \|w_i\|$. If $b_i = 0$, the hyperplane cuts through the origin. In high-dimensional vector spaces (e.g., $d = 4096$), volume concentration phenomena dictate that most of the mass of a standard Gaussian distribution lies in a thin shell away from the origin. If all decision boundaries are forced to pass through the origin, they slice this shell symmetrically, significantly constraining the types of geometric partitions the network can form.

By introducing a bias $b \neq 0$, the network can shift these hyperplanes away from the origin, allowing it to carve out localized polytopes (bounded regions) or select specific segments of the representational shell. A positive bias $b_i > 0$ makes a ReLU unit "active by default" near the origin, requiring a strong opposing signal in x to deactivate it. A negative bias $b_i < 0$ makes the unit "inactive by default," acting as a threshold that the signal $w_i \cdot x$ must overcome. This thresholding capability is fundamental to mechanism design in neural networks, allowing layers to act as filters that only pass signals exceeding a certain magnitude or matching a specific pattern.

Despite the theoretical necessity of biases for general affine maps, some modern Transformer implementations omit biases in specific layers (e.g., in the Q, K, V projections of attention or even the FFN). This design choice relies on the presence of normalization layers (like LayerNorm or RMSNorm) effectively re-centering data, or on the observation that for extremely high-dimensional parameters, the degrees of freedom in W alone are sufficient to approximate the necessary transformations. However, the standard mathematical formulation assumes the general affine case, and the omission of bias is best understood as a regularizing constraint—forcing the learned map to be purely linear—rather than a simplification of the underlying geometry.

The affine transformation provides the capacity to rotate, scale, and shift the data manifold, but it cannot change the topology or curvature of the representation. A composition of L affine maps is simply another affine map:

$$W_L(\dots(W_2(W_1x + b_1) + b_2)\dots) + b_L = W'x + b'.$$

Without nonlinearities, a deep network collapses into a single linear operator, rendering depth redundant. The pointwise nonlinearity $\sigma : \mathbb{R} \to \mathbb{R}$ is the essential component that breaks this linearity, enabling the approximation of complex functions.

In the context of the Transformer, these nonlinearities are applied element-wise to the output of the affine map. If $Z = HW^\top + \mathbf{1}b^\top$, the activation is $A = \sigma(Z)$, where σ is applied to every entry Z_{tj}. This element-wise application implies that the nonlinearity acts independently on each coordinate of the projected space. The choice of basis for this projection is determined by the rows of W. Thus, the "features" detected by the layer are defined by the linear forms $w_j \cdot x$, and the nonlinearity dictates how the presence of that feature is converted into an activation.

The Rectified Linear Unit (ReLU), defined as $\text{ReLU}(z) = \max(0, z)$, was the standard choice in early deep learning and the original Transformer. Its geometric interpretation is clean: it folds the vector space. For a single neuron, the input space is divided into two regions by the hyperplane $w \cdot x + b = 0$. On one side, the function is identity (linear); on the other, it is zero (flat). When combining d_{ff} such neurons, the input space $\mathbb{R}^{d_{\mathrm{in}}}$ is partitioned into a complex arrangement of convex polytopes. Within each polytope, the network behaves as a linear function. The global function computed by a ReLU network is continuous and piecewise linear. This implies that the "curvature" of the learned manifold is concentrated entirely at the boundaries of these polytopes (the "kinks").

More recently, smooth nonlinearities like the Gaussian Error Linear Unit (GELU) have replaced ReLU in state-of-the-art LLMs (e.g., GPT-3, Llama). The GELU is defined as:

$$\mathrm{GELU}(x) = x\Phi(x),$$

where $\Phi(x)$ is the cumulative distribution function of the standard normal distribution. Unlike ReLU, GELU is smooth (differentiable everywhere) and non-monotonic. For $x \gg 0$, $\mathrm{GELU}(x) \approx x$. For $x \ll 0$, $\mathrm{GELU}(x) \to 0$. However, in the transition region around zero, GELU is slightly negative before turning positive.

The shift to smooth activations like GELU, Swish, or SiLU is motivated by optimization dynamics and the "probabilistic" interpretation of activation. GELU can be viewed as an expectation: it is the expected output of a deterministic input x multiplied by a stochastic 0-1 mask, where the mask probability depends on x itself. Smoothness ensures that gradients are well-defined and non-zero everywhere (except at limits), avoiding the "dead ReLU" problem where a neuron stuck in the negative region receives zero gradient and stops learning. Furthermore, the local curvature introduced by smooth functions allows the network to approximate higher-order terms in the Taylor expansion of the target function more efficiently than a piecewise linear approximation could with the same number of parameters.

When we compose the affine map with the nonlinearity, we define the "Feed-Forward Network" (FFN) block, typically structured as an expansion followed by a contraction. A common configuration in Transformers is $d_{\text{model}} \to 4d_{\text{model}} \to d_{\text{model}}$. Mathematically, this block computes:

$$\text{FFN}(x) = W_2(\sigma(W_1 x + b_1)) + b_2,$$

where $W_1 \in \mathbb{R}^{d_{\text{ff}} \times d_{\text{model}}}$ and $W_2 \in \mathbb{R}^{d_{\text{model}} \times d_{\text{ff}}}$. The intermediate dimension d_{ff} is usually much larger than the embedding dimension d_{model} (e.g., $4\times$). This expansion is critical. By projecting the data into a higher-dimensional space, the affine-plus-nonlinearity transformation can disentangle representations that were linearly inseparable in the lower dimension. The nonlinearity σ acts effectively in this overcomplete basis, activating sparse subsets of the expanded features. The second matrix W_2 then projects these activated features back down to the model dimension, aggregating the processed information.

This structure—expansion, element-wise nonlinearity, contraction—resembles a key-value memory network where the keys are the rows of W_1 and the values are the rows of $W_2^\top$. The input x is compared against all keys via dot products ($W_1 x$). The bias and nonlinearity determine the "match strength" or activation level for each key. The output is then a weighted sum of the value vectors, weighted by these activation levels. This interpretation links the abstract affine geometry back to the functional goal of the layer: retrieving and composing distributed knowledge stored in the weights.

We must also consider the operator norms of these affine maps. The spectral norm $||W||_2$ (the largest singular value) dictates the maximum expansion factor of the Euclidean length of any input vector: $||Wx||_2 \leq ||W||_2 ||x||_2$. If the spectral norms of the weight matrices are significantly greater than 1, signal magnitudes can grow exponentially with depth, leading to instability. Conversely, if they are less than 1, signals may vanish. The combination of W and the nonlinearity σ determines the Lipschitz constant of the layer. Since common activation functions like ReLU and GELU have Lipschitz constants close to 1 (ReLU is exactly 1, GELU is slightly larger than 1 in its active range), the stability of the network is largely controlled by the singular values of W. This motivates initialization schemes (like Xavier or Kaiming initialization) that set the initial variance of W such that the operator norm is close to 1, preserving the variance of activations across layers.

In summary, the affine layers of the Transformer are not passive conduits but active geometric operators. The matrix W performs basis change and dimensionality scaling; the bias b provides the necessary translational freedom to position decision boundaries away from the origin; and the element-wise nonlinearity σ enables the segmentation of the vector space into distinct functional regions. The interplay of these components allows the FFN to act as a universal approximator on the manifold of token embeddings, performing the dense logical and associative processing required for language understanding. The "shape discipline" of maintaining consistent vector, matrix, and batch dimensions ensures that these operations remain efficient and parallelizable on modern hardware, forming the robust computational substrate of the model.

Chapter 3

The Mechanics of Self-Attention

Self-attention is the Transformer's "token-mixing" operator: a single, carefully parameterized matrix pipeline that converts a sequence of vectors into a new sequence where each position is a content-dependent weighted combination of other positions. This chapter treats attention as a precise mathematical object-tracking shapes, invariances, and stability-so later chapters can treat the Transformer block as a composition of well-understood maps rather than a black box.

3.1. Scaled Dot-Product Attention as a Differentiable Content-Addressing Operator

At the core of the Transformer architecture lies the scaled dot-product attention mechanism, a differentiable operation that maps a query and a set of key-value pairs to an output. Unlike fixed operational layers such as convolutions, which apply a static kernel over a local neighborhood, attention constructs a dynamic kernel based on the input content itself. This mechanism allows the model to retrieve information from arbitrary positions in a sequence based on computed relevance rather than fixed geometric proximity.

We begin by formally defining the input state. Let $X \in \mathbb{R}^{T \times d_{model}}$ represent a sequence of T input vectors, where the t-th row $x_t \in \mathbb{R}^{d_{model}}$ corresponds to the representation of the token at position t. In a self-attention layer, this input matrix X projects into three distinct subspaces: the Query (Q), Key (K), and Value (V). These projections are parameterized by three learned weight matrices:

$$W_Q \in \mathbb{R}^{d_{model} \times d_k}$$
$$W_K \in \mathbb{R}^{d_{model} \times d_k}$$
$$W_V \in \mathbb{R}^{d_{model} \times d_v}$$

The projections are computed via matrix multiplication:

$$Q = XW_Q, \quad K = XW_K, \quad V = XW_V$$

Resulting in matrices $Q \in \mathbb{R}^{T \times d_k}$, $K \in \mathbb{R}^{T \times d_k}$, and $V \in \mathbb{R}^{T \times d_v}$. It is crucial to distinguish between d_{model}, the width of the residual stream, and d_k, d_v, the internal dimensions of the attention head. While d_k and d_v are often set equal (e.g., $d_k = d_v = d_{model}/h$ in multi-head settings), they serve distinct roles. The dimension d_k determines the geometry of the similarity space where queries and keys interact, while d_v determines the dimensionality of the retrieved content.

The fundamental operation of attention is content-based addressing. For every query vector q_t (the t-th row of Q), we compute a relevance score against all key vectors k_s (the rows of K). The canonical measure of similarity in this latent space is the dot product. We form the pre-softmax score matrix $S \in \mathbb{R}^{T \times T}$ by computing the dot products between all pairs of queries and keys:

$$S = QK^\top$$

Here, the entry $S_{t,s} = q_t \cdot k_s$ represents the unnormalized logit indicating the compatibility between the query at position t and the key at position s.

Geometric interpretation of the score matrix reveals why this operation scales quadratically with sequence length. The matrix multiplication $QK^\top$ involves T^2 dot products, each of dimension d_k. The resulting $T \times T$ interaction graph is dense; every token attends to every other token (prior to masking). This density is the source of the Transformer's global receptive field, allowing it to model dependencies regardless of distance, but it also imposes the $O(T^2)$ memory and compute bottleneck that characterizes the architecture.

A critical, often under-analyzed component of the attention formula is the scaling factor $1/\sqrt{d_k}$. This term is not merely a heuristic for numerical stability; it is a variance-stabilizing normalization essential for gradient-based optimization in high-dimensional spaces.

Consider the dot product of two random vectors $q, k \in \mathbb{R}^{d_k}$. Assume that the components q_i and k_i are independent and identically distributed random variables with mean 0 and variance σ^2. The dot product is given by:

$$u = q \cdot k = \sum_{i=1}^{d_k} q_i k_i$$

To analyze the distribution of u, we examine its mean and variance. The expectation is:

$$\mathbb{E}[u] = \sum_{i=1}^{d_k} \mathbb{E}[q_i]\mathbb{E}[k_i] = 0$$

The variance, assuming independence between q and k, is the sum of the variances of the terms:

$$\text{Var}(u) = \sum_{i=1}^{d_k} \text{Var}(q_i k_i)$$

For independent variables, $\text{Var}(XY) = \mathbb{E}[X^2]\mathbb{E}[Y^2] - (\mathbb{E}[X]\mathbb{E}[Y])^2$. With zero mean and variance σ^2:

$$\text{Var}(q_i k_i) = \sigma^2 \cdot \sigma^2 - 0 = \sigma^4$$

Thus, the variance of the dot product scales linearly with the dimension d_k:

$$\text{Var}(u) = d_k \sigma^4$$

The standard deviation of the dot product is $\sigma^2\sqrt{d_k}$. As d_k increases, the range of values in the score matrix S expands. For a typical value like $d_k = 64$, the standard deviation increases by a factor of 8; for $d_k = 128$, by roughly 11.3.

This growth in magnitude has profound consequences for the subsequent softmax operation. The softmax function $\sigma(\mathbf{z})_i = \frac{e^{z_i}}{\sum_j e^{z_j}}$ is sensitive to the scale of its inputs. When the input logits z_i have large magnitudes, the exponential function magnifies the differences. The largest logit dominates the denominator, causing the output distribution to approach a "one-hot" vector where the probability mass is concentrated on a single index.

In this saturation regime, the local gradients of the softmax function vanish. The Jacobian of the softmax function for a vector $\mathbf{z}$ is given by:

$$\frac{\partial \sigma_i}{\partial z_j} = \sigma_i(\delta_{ij} - \sigma_j)$$

If the output distribution is extremely peaked (e.g., $\sigma_k \approx 1$ for some k and $\sigma_i \approx 0$ for $i \neq k$), the diagonal term $\sigma_k(1 - \sigma_k)$ approaches 0, and the off-diagonal terms $-\sigma_i\sigma_j$ also approach 0. Consequently, the gradient signal backpropagating through the attention layer becomes negligible, halting learning.

To counteract this, we scale the dot products by $1/\sqrt{d_k}$. Under the same statistical assumptions, the variance of the scaled scalar $\frac{u}{\sqrt{d_k}}$ becomes:

$$\mathrm{Var}\left(\frac{u}{\sqrt{d_k}}\right) = \frac{1}{d_k}\mathrm{Var}(u) = \frac{1}{d_k}(d_k\sigma^4) = \sigma^4$$

By enforcing that the variance of the logits is independent of the dimension d_k, we ensure that the softmax inputs remain within a regime where the gradients are well-behaved, regardless of the head size. This scaling effectively decouples the optimization dynamics from the choice of hyperparameters.

We define the attention weights (or attention probabilities) $A \in \mathbb{R}^{T\times T}$ by applying the softmax function row-wise to the scaled scores:

$$A = \mathrm{softmax}\left(\frac{QK^\top}{\sqrt{d_k}}\right)$$

Each row $A_{t,:}$ represents a discrete probability distribution over the sequence positions $\{1, \ldots, T\}$, summing to 1. Specifically, $A_{t,s}$ is the weight assigned to the value at position s when constructing the output for position t.

The final output of the attention mechanism, $Y \in \mathbb{R}^{T\times d_v}$, is computed as the weighted sum of the value vectors V:

$$Y = AV$$

This operation can be interpreted as a soft lookup in an associative memory. The keys define the addresses, the queries define the search intent, and the values constitute the content stored at those addresses. Because A is row-stochastic, each row y_t of Y is a convex combination of the rows of V. The vector y_t aggregates information from the entire sequence, weighted by relevance.

From a tensor perspective, the shapes align as follows:

- Q: (T, d_k)

- $K^\top$: (d_k, T)
- $QK^\top$: (T, T) — The pairwise score matrix.
- A: (T, T) — The stochastic attention matrix.
- V: (T, d_v)
- AV: (T, d_v) — The sequence of context vectors.

To solidify these concepts, we provide a reference implementation using PyTorch. This implementation highlights the explicit manipulation of dimensions and the application of scaling.

```
import torch
import torch.nn.functional as F
import math

def scaled_dot_product_attention(query, key, value, mask=None):
    """
    Compute 'Scaled Dot Product Attention'.

    Args:
        query: Tensor of shape (batch, heads, seq_len_q, d_k)
        key:   Tensor of shape (batch, heads, seq_len_k, d_k)
        value: Tensor of shape (batch, heads, seq_len_k, d_v)
        mask:  Optional boolean tensor (batch, 1, seq_len_q,
     seq_len_k)
               False indicates positions to ignore (mask out).

    Returns:
        context: Tensor of shape (batch, heads, seq_len_q, d_v)
        attn:    Attention weights of shape (batch, heads, seq_len_q,
      seq_len_k)
    """
    d_k = query.size(-1)

    # 1. Compute unnormalized scores (logits)
    # Shape: (batch, heads, seq_len_q, seq_len_k)
    scores = torch.matmul(query, key.transpose(-2, -1))

    # 2. Apply scaling factor
    scores = scores / math.sqrt(d_k)

    # 3. Apply mask (if provided)
    if mask is not None:
```

```
            # Fill masked positions with a very large negative number
            scores = scores.masked_fill(mask == 0, -1e9)

        # 4. Compute attention probabilities
        attn_weights = F.softmax(scores, dim=-1)

        # 5. Weighted sum of values
        # Shape: (batch, heads, seq_len_q, d_v)
        context = torch.matmul(attn_weights, value)

        return context, attn_weights

# Example Usage
# Batch=1, Heads=1, T=4, d_k=8, d_v=8
d_k = 8
q = torch.randn(1, 1, 4, d_k)
k = torch.randn(1, 1, 4, d_k)
v = torch.randn(1, 1, 4, d_k)

out, weights = scaled_dot_product_attention(q, k, v)
```

```
Input shapes:
Q: torch.Size([1, 1, 4, 8])
K: torch.Size([1, 1, 4, 8])
V: torch.Size([1, 1, 4, 8])

Output shape:
Context: torch.Size([1, 1, 4, 8])
Weights sum (check row stochasticity): tensor([1., 1., 1., 1.])
```

The implementation explicitly handles the batch and head dimensions, which we treated implicitly in the single-matrix derivation (X, Q, K, V). In practice, the operation broadcasts over the batch dimension and operates independently over the head dimension, a detail that facilitates the parallelization of multi-head attention (discussed in subsequent sections).

The output Y is a refined representation of the input sequence. While the feed-forward networks (discussed in Chapter 4) process each token in isolation to extract static features, the self-attention layer mixes information across the time axis. It transforms the sequence X into a new sequence Y where y_t contains features not just from x_t, but from any x_s deemed relevant by the projection W_Q and W_K. This differentiability of the retrieval process-where the "address" is a continuous vector rather than a discrete pointer-enables the model to learn complex routing strategies via backpropagation.

This formulation defines the "unmasked" attention mechanism. In the context of autoregressive language modeling, however, this mechanism allows information to flow from future tokens ($s > t$) to the current position t, violating causality. The necessary modification to this operator to enforce temporal order—causal masking—builds directly upon the score matrix S derived here.

3.2. Autoregressive Consistency: Causal Masking as Constrained Normalization

The fundamental contract of an autoregressive sequence model is the decomposition of the joint probability distribution into a product of conditional probabilities, $P(X) = \prod_{t=1}^{T} P(x_t \mid x_{<t})$. When training such models using the teacher-forcing paradigm, the entire sequence X is available to the model simultaneously. While this enables the massive parallelism characteristic of Transformers, it introduces a risk: without explicit structural constraints, the representation at position t could inadvertently incorporate information from positions $s > t$, violating the causal dependency required for valid probability factorization. In Recurrent Neural Networks, causality is enforced by the sequential nature of the computation. In self-attention, which connects every position to every other position via the score matrix S, causality must be enforced by modifying the attention operator itself.

We formalize this constraint not as a post-hoc filtering of outputs, but as a domain restriction on the normalization function. Recall that the standard attention mechanism computes a dense score matrix $S =$

$QK^\top/\sqrt{d_k} \in \mathbb{R}^{T\times T}$, where the entry $S_{t,s}$ represents the raw affinity between query t and key s. The subsequent softmax operation normalizes these scores into a probability distribution over the sequence length T. To enforce autoregressive consistency, we must ensure that the probability mass assigned to any future token is exactly zero:

$$A_{t,s} = P(\text{attend to } s \mid \text{at } t) = 0 \quad \forall s > t.$$

Furthermore, the remaining probabilities must define a valid distribution over the history, meaning $\sum_{s=1}^{t} A_{t,s} = 1$. This implies that the partition function (the denominator of the softmax) for row t must sum only over the indices $\{1, \ldots, t\}$.

This is achieved efficiently by additive causal masking. We define a constant mask matrix $M \in \mathbb{R}^{T\times T}$ such that:

$$M_{t,s} = \begin{cases} 0 & \text{if } s \leq t \\ -\infty & \text{if } s > t \end{cases}$$

The masked score matrix is given by $\tilde{S} = S + M$. When the softmax function is applied to $\tilde{S}$, the exponential of the upper-triangular entries becomes zero ($\lim_{x\to-\infty} e^x = 0$). Consequently, these entries contribute nothing to the partition function $Z_t = \sum_{k=1}^{T} e^{\tilde{S}_{t,k}} = \sum_{k=1}^{t} e^{S_{t,k}}$, effectively restricting the normalization to the causal prefix.

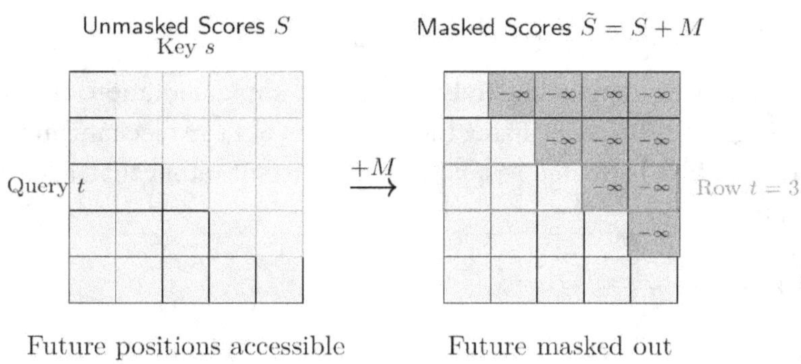

The additive formulation is mathematically preferred over multiplicative masking (zeroing out probabilities after softmax) because it operates in the logit domain. Zeroing out probabilities post-softmax would violate the summation constraint ($\sum A_{t,:} < 1$) and require a second renormalization step. By modifying the logits directly, the softmax naturally yields a valid distribution over the available history.

In practical implementations, the theoretical value $-\infty$ is replaced by a large negative finite constant, often denoted as M_{neg}. The magnitude of this constant requires careful selection based on the numerical precision of the floating-point format used (e.g., IEEE 754 float32 or bfloat16). If M_{neg} is too small in magnitude, such as -100, and the unmasked logits are large (e.g., $+100$ due to insufficient scaling), the "masked" probability $\exp(-100 + 100) = 1$ would not vanish. Conversely, if M_{neg} is too large, such as -10^{30}, it may trigger raw Not-a-Number (NaN) errors during gradient computation or underflow to zero in a way that destabilizes the softmax backward pass.

For standard 32-bit floating point arithmetic, a value of -10^9 is commonly used, as $\exp(-10^9)$ is effectively zero within machine precision. In 16-bit mixed-precision training, the dynamic range is significantly smaller; here, the mask value must be chosen to be the minimum representable finite number (e.g., approximately $-65{,}504$ for float16) to avoid overflow to $-$Inf, which produces NaNs when subtracted or processed in fused kernels.

The following listing illustrates a robust implementation of causal masking in PyTorch, utilizing the `register_buffer` mechanism to ensure the mask is treated as a persistent constant rather than a learned parameter.

```
import torch
import torch.nn as nn

class CausalSelfAttention(nn.Module):
    def __init__(self, d_model, block_size):
        super().__init__()
```

```
        # Create a lower-triangular matrix of ones
        # shape: (block_size, block_size)
        mask = torch.tril(torch.ones(block_size, block_size))

        # Register as buffer to save in state_dict but not update via
     optim
        # We store it as a boolean mask for flexibility
        self.register_buffer('bias', mask.view(1, 1, block_size,
    block_size))

    def forward(self, q, k, v):
        # q, k, v shape: (B, H, T, C)
        att = (q @ k.transpose(-2, -1)) * (1.0 / math.sqrt(k.size(-1)
    ))

        # Apply mask: fill positions where mask is 0 with -inf
        # slicing [:T, :T] handles sequences shorter than block_size
        T = q.size(2)
        att = att.masked_fill(self.bias[:,:,:T,:T] == 0, float('-inf
    '))

        y = torch.nn.functional.softmax(att, dim=-1)
        return y @ v
```

This implementation detail-using `masked_fill` with `float('-inf')`-relies on the underlying softmax kernel correctly handling infinities. Modern deep learning frameworks ensure that softmax(x) returns 0 for entries where $x_i = -\infty$ and correctly normalizes the remaining entries.

From an optimization perspective, causal masking does more than enforce logic; it modifies the optimization landscape by altering the flow of gradients. During the backward pass, the gradient of the loss function L with respect to the score matrix S is computed via the Jacobian of the softmax. Let $y = \text{softmax}(S + M)$. The Jacobian entries are $J_{ij} = y_i(\delta_{ij} - y_j)$. For any masked position (t, s) where $s > t$, the forward probability $y_{t,s}$ is exactly zero. Consequently, the partial derivative of the loss with respect to the score $S_{t,s}$ at that forbidden position is:

$$\frac{\partial L}{\partial S_{t,s}} = \sum_k \frac{\partial L}{\partial y_{t,k}} \frac{\partial y_{t,k}}{\partial S_{t,s}} = 0.$$

This zero gradient confirms that the mask effectively detaches the forbidden connections from the computational graph. The model receives no signal to increase or decrease the affinity for future tokens because those affinities never influence the output. This sparsity is beneficial: it concentrates the learning signal entirely on the valid historical relationships, preventing the optimizer from wasting effort on suppressing future-looking dependencies that are structurally impossible.

However, the "constrained normalization" view reveals a subtle edge case in attention mechanics. Because the softmax must sum to 1 over the valid set $\{1, \ldots, t\}$, the attention mechanism is forced to assign probability mass somewhere within the history. For the very first token ($t = 1$), the valid set is a singleton $\{1\}$. Therefore, $A_{1,1} = 1$ regardless of the values of Q and K. The first position effectively attends entirely to itself, acting as a pass-through for its own embedding (and positional encoding). As t increases, the model gains the flexibility to distribute mass, but it cannot choose to "attend to nothing." Even if the current context is entirely irrelevant to the query, the mechanism must allocate 100% of the attention weight among the available history. This property, inherent to the softmax geometry, motivates the inclusion of *sink tokens* or specific bias terms in some advanced architectures to provide a "null" attention target, though the standard Transformer relies on the learned projections W_V to suppress the magnitude of the output vector if necessary.

The application of the causal mask transforms the self-attention mechanism from a fully connected graph into a Directed Acyclic Graph (DAG) over the sequence positions. While the matrix multiplication $QK^\top$ computes T^2 interactions, the mask nullifies strictly fewer than half of them ($T(T-1)/2$). This retains the $O(T^2)$ computational complexity, as the dense tensor operations are performed before masking. Optimization of this quadratic cost, particularly for inference (where T grows one step at a time), relies on KV-caching and specialized kernels, but the theoretical definition of the operator remains rooted in this dense, masked matrix formulation. The mask ensures that the parallel computation of attention across all t during training is mathematically equivalent to T sequential steps of inference, satisfying the autoregressive consistency required for correct probabilistic modeling.

3.3. Multi-Head Self-Attention: Block-Structured Projections, Subspace Factorization, and Expressivity

While the scaled dot-product attention mechanism provides a differentiable means of routing information based on content, applying it as a single monolithic operator over the entire model dimension d_{model} imposes significant constraints on representational capacity. A single attention head computes a single probability distribution over the context for each query. Consequently, a specific position t can focus on only one set of semantic associates at a time. If a token needs to attend to both its syntactic dependency (e.g., a verb attending to its subject) and its semantic antecedent (e.g., a pronoun attending to a named entity) simultaneously, a single probability mass function forces a compromise—an average location that may represent neither relationship effectively.

Multi-Head Attention (MHA) resolves this by factorizing the attention mechanism into h parallel subspaces. Rather than a single global view of the sequence, the model projects the input X into h distinct query-key-value spaces, performs attention independently within each, and recombines the results. This section formalizes MHA not merely as a parallelization trick, but as a structured rank decomposition of the mixing operator, examining its tensor geometry, algebraic properties, and the resulting implications for model expressivity.

Mathematical Formulation of the Multi-Head Operator

Let $X \in \mathbb{R}^{T \times d_{\text{model}}}$ denote the input sequence representation. We define h attention heads, indexed by $i \in \{1, \ldots, h\}$. For each head, we introduce three distinct linear projection matrices:

$$W_Q^{(i)} \in \mathbb{R}^{d_{\text{model}} \times d_k}$$
$$W_K^{(i)} \in \mathbb{R}^{d_{\text{model}} \times d_k}$$
$$W_V^{(i)} \in \mathbb{R}^{d_{\text{model}} \times d_v}$$

In standard implementations, such as the original Transformer architecture, the head dimension is often chosen such that $d_k = d_v = d_{\text{model}}/h$, preserving the total parameter count relative to a single-head projection of full dimension. However, this is a design choice, not a strict mathematical requirement; d_k and d_v are free hyperparameters governing the rank of the attention mechanism.

The projection step maps the high-dimensional input features into lower-dimensional subspaces specific to each head:

$$Q^{(i)} = XW_Q^{(i)}, \quad K^{(i)} = XW_K^{(i)}, \quad V^{(i)} = XW_V^{(i)}$$

Each head then computes its own attention output, $H^{(i)} \in \mathbb{R}^{T \times d_v}$, using the scaled dot-product mechanism derived previously. Note that the scaling factor becomes $1/\sqrt{d_k}$ specific to the head dimension, ensuring variance control within the subspace:

$$H^{(i)} = \text{softmax}\left(\frac{Q^{(i)}(K^{(i)})^\top}{\sqrt{d_k}}\right)V^{(i)}$$

Crucially, the softmax is computed independently for each head. This allows the model to capture up to h distinct alignment patterns for every position t. One head might attend to the previous token (local context), while another attends to a specific token globally (long-range dependency).

The outputs of the h heads are combined via concatenation followed by a final linear transformation. Let $\text{Concat}(H^{(1)}, \dots, H^{(h)}) \in \mathbb{R}^{T \times (h \cdot d_v)}$ denote the assembled multi-head representation. The final output Y is produced by the output projection matrix $W_O \in \mathbb{R}^{(h \cdot d_v) \times d_{\text{model}}}$:

$$Y = \text{Concat}(H^{(1)}, \dots, H^{(h)}) W_O$$

This linear projection W_O serves a dual purpose: it mixes the information retrieved from different heads—allowing features extracted in separate subspaces to interact—and projects the concatenated vector back to the residual stream dimension d_{model}.

The Additive Decomposition View

While the concatenation formulation is standard for implementation, an additive view offers superior theoretical insight into how heads contribute to the final representation. We can block-partition the output matrix W_O into h sub-matrices $W_O^{(i)} \in \mathbb{R}^{d_v \times d_{\text{model}}}$ corresponding to the slice of the concatenated input associated with head i:

$$W_O = \begin{bmatrix} W_O^{(1)} \\ W_O^{(2)} \\ \vdots \\ W_O^{(h)} \end{bmatrix}$$

Substituting this into the output equation reveals that multi-head attention is mathematically equivalent to a sum of independent linear operators:

$$Y = \sum_{i=1}^{h} H^{(i)} W_O^{(i)} = \sum_{i=1}^{h} \left[\text{softmax}\left(\frac{X W_Q^{(i)} (X W_K^{(i)})^\top}{\sqrt{d_k}} \right) X W_V^{(i)} \right] W_O^{(i)}$$

This decomposition highlights that MHA operates as an ensemble of updates. Each term in the summation represents a full-rank update (in the d_{model} space) derived from a low-rank attention mechanism. The matrix $W_V^{(i)}$ projects inputs into a value subspace, the attention weights mix them, and $W_O^{(i)}$ projects them back. This structure suggests that W_O is not merely a dimension-restoring matrix but a weighting mechanism that determines the relative importance of each head's output to the construction of the next layer's input.

Tensor Geometry and Parallel Implementation

In practice, computing the projections and attention scores sequentially for each head is inefficient. Modern deep learning frameworks utilize the block-diagonal structure of the operations to vectorize computation. The projections $W_Q^{(i)}$ for all i are often fused into a single matrix $W_Q^{\text{all}} \in \mathbb{R}^{d_{\text{model}} \times (h \cdot d_k)}$.

The forward pass involves tensor reshaping (view) and permutation (transpose) operations to treat the head dimension as a batch dimension. This enables the use of highly optimized Batch Matrix Multiply (BMM) kernels.

```
import torch
import torch.nn as nn
import math

class MultiHeadAttention(nn.Module):
    def __init__(self, d_model, num_heads):
        super().__init__()
        assert d_model % num_heads == 0

        self.d_k = d_model // num_heads
        self.h = num_heads

        # Fused projections for Q, K, V
        # Shape: (d_model, 3 * d_model)
        self.w_qkv = nn.Linear(d_model, 3 * d_model)
        self.w_o = nn.Linear(d_model, d_model)

    def forward(self, x, mask=None):
        B, T, D = x.size()

        # 1. Fused projection
        # qkv: [B, T, 3 * h * d_k]
        qkv = self.w_qkv(x)

        # 2. Reshape and Transpose to separate heads
```

```
        # Split into (Q, K, V) and reshape to [B, h, T, d_k]
        qkv = qkv.reshape(B, T, 3, self.h, self.d_k)
        qkv = qkv.permute(2, 0, 3, 1, 4)
        q, k, v = qkv[0], qkv[1], qkv[2]

        # 3. Scaled Dot-Product Attention (Batch Matrix Multiply)
        # scores: [B, h, T, T]
        scores = (q @ k.transpose(-2, -1)) / math.sqrt(self.d_k)

        if mask is not None:
            # Mask expansion handled externally or broadcast here
            scores = scores.masked_fill(mask == 0, -1e9)

        attn = torch.softmax(scores, dim=-1)

        # 4. Weighted Sum and Output Projection
        # context: [B, h, T, d_k] -> [B, T, h, d_k] -> [B, T, d_model
]
        context = attn @ v
        context = context.transpose(1, 2).reshape(B, T, D)

        return self.w_o(context)
```

This implementation underscores the structural rigidity of the mechanism—the interaction between heads is strictly prohibited until the final mixing at `w_o`. The `permute` operation effectively isolates the subspaces, ensuring that the dot product $q \cdot k$ occurs only between coordinates belonging to the same head i.

Subspace Factorization and Metric Learning

The “multi-head” terminology often obscures the geometric reality: MHA is a mechanism for learning multiple distinct similarity metrics over the same data. The scalar attention score for head i between positions t and s is:

$$\text{Score}_{t,s}^{(i)} = \frac{1}{\sqrt{d_k}} x_t W_Q^{(i)} (x_s W_K^{(i)})^\top = \frac{1}{\sqrt{d_k}} x_t \left(W_Q^{(i)} {W_K^{(i)}}^\top \right) x_s^\top$$

Let $M^{(i)} = W_Q^{(i)} {W_K^{(i)}}^\top$. The matrix $M^{(i)}$ defines a bilinear form on the input space $\mathbb{R}^{d_{\text{model}}}$. MHA, therefore, learns h such bilinear forms simultaneously. Since $W_Q^{(i)}$ and $W_K^{(i)}$ are generally low-rank ($d_k < d_{\text{model}}$),

$M^{(i)}$ is a low-rank matrix that projects x_t and x_s onto a subspace before comparing them.

This implies that two vectors x_t and x_s can be "similar" in the subspace of head 1 (e.g., sharing a syntactic role) while being "dissimilar" or orthogonal in the subspace of head 2 (e.g., differing in semantic topic). By maintaining disjoint parameter sets for these projections, the Transformer avoids the interference that would occur if a single metric attempted to capture all relevant linguistic and structural relationships simultaneously.

Invariances and Parameter Redundancy

The factorization into Q, K, V projections introduces specific symmetries (invariances) in the parameter space. Understanding these is vital for analyzing model optimization and pruning.

Consider the interaction within a single head. The attention scores depend on the product $QK^\top$. If we insert an invertible matrix $R \in \mathbb{R}^{d_k \times d_k}$ and its inverse, the scores remain unchanged:

$$(QR)(K(R^{-1})^\top)^\top = QRR^{-1}K^\top = QK^\top$$

This means the specific basis chosen for the query/key subspace d_k is not identifiable; only the subspace itself and the geometry defined by the product $W_Q^{(i)} W_K^{(i)\top}$ matter. Specifically, applying a rotation (orthogonal matrix) to both $W_Q^{(i)}$ and $W_K^{(i)}$ leaves the attention pattern $A^{(i)}$ entirely invariant.

Furthermore, there is an interplay between $W_V^{(i)}$ and $W_O^{(i)}$. Recall that the output contribution of head i is $(A^{(i)} X W_V^{(i)}) W_O^{(i)}$. If we scale $W_V^{(i)}$ by a scalar α and $W_O^{(i)}$ by $1/\alpha$, the output is unchanged. More generally, linear transformations can migrate between the value projection and the output projection, limited only by the rank constraints of d_v. This redundancy suggests that the effective capacity of the model may be lower than a naive parameter count implies, and it helps explain the success of head pruning techniques where entire heads are removed post-training with minimal performance degradation.

Expressivity and Limitations

The expressivity of Multi-Head Attention is bounded by two primary factors: the rank of the subspaces (d_k) and the row-stochastic constraint of the softmax.

- **Rank Bottleneck:** If d_k is too small (low dimension per head), the projection $XW_Q^{(i)}$ may collapse distinguishable input vectors into the same query representation, causing "attention confusion" where the head cannot resolve differences between distinct keys. While the total dimension $h \cdot d_k$ usually equals d_{model}, the fragmentation into disjoint heads means we trade a single high-fidelity comparison for h lower-fidelity comparisons. This trade-off is generally favorable for natural language, which involves multifaceted discrete relationships, but can be a limitation in continuous modalities requiring high-precision regression.
- **The Softmax Constraint:** Each head produces a convex combination of value vectors. The output of head i lies strictly within the convex hull of the value vectors $\{v_1^{(i)}, \dots, v_T^{(i)}\}$. MHA cannot perform arbitrary nonlinear transformations of the input; it can only mix them. Feature extraction (creating new features not present in the linear span of inputs) is principally the domain of the feed-forward networks (discussed in Chapter 4). MHA is strictly a routing and mixing mechanism.

Recent research has highlighted the phenomenon of "head collapse," where multiple heads in a trained Transformer converge to identical or highly correlated attention patterns. While this suggests inefficiency, it also provides robustness-redundant heads may act as a backup mechanism or smooth the optimization landscape by providing an over-parameterized search space during early training. When heads do differentiate, they typically specialize into functional groups: "positional heads" that attend to relative offsets (e.g., $t-1$), "syntactic heads" that track dependency parses, and "rare word heads" that point to low-frequency tokens. This emergence of functional specialization confirms that the block-structured factorization of MHA successfully decomposes the complex task of sequence modeling into parallel, tractable sub-problems.

Chapter 4

Transformer Architecture and Signal Propagation

A decoder-only Transformer is not "attention plus a couple of layers"- it is a carefully engineered composition whose algebra is tuned for stable propagation at scale. In this chapter we model each sublayer as an operator on residual streams, analyze the geometry induced by normalization and residual addition, and develop a block-level view that explains why modern LLMs can be both extremely deep and trainable.

4.1. Position-Wise Feed-Forward Networks as Channel Mixing Operators

The Transformer architecture partitions its computational responsibilities along two orthogonal axes of the input tensor: the sequence length

T and the feature dimension d. While the self-attention mechanism handles interactions across the sequence length—mixing information between tokens—the Position-Wise Feed-Forward Network (FFN) operates exclusively along the feature dimension. This separation of concerns allows us to formalize the FFN not merely as a generic neural network layer, but as a specific class of nonlinear operators that perform "channel mixing" without "token mixing."

Let the residual stream at a given layer be represented by a matrix $X \in \mathbb{R}^{T\times d}$, where T is the number of tokens and d is the model width. The defining characteristic of the position-wise FFN is that it factorizes across the token index t. Formally, the mapping $\Phi : \mathbb{R}^{T\times d} \to \mathbb{R}^{T\times d}$ applies a single function $f : \mathbb{R}^d \to \mathbb{R}^d$ identically to each row vector X_t:

$$(\Phi(X))_t = f(X_t) \quad \text{for } t = 1, \ldots, T.$$

Because f does not depend on $X_{t'}$ for $t' \neq t$, the FFN cannot move information between positions. Its role is strictly to process the internal representation of each token, transforming the features synthesized by previous attention layers.

The canonical realization of f, as introduced in the original Transformer formulation, consists of two affine transformations separated by a pointwise nonlinearity σ:

$$f(x) = W_2\sigma(W_1x + b_1) + b_2,$$

where $W_1 \in \mathbb{R}^{d_{ff}\times d}$, $W_2 \in \mathbb{R}^{d\times d_{ff}}$, and b_1, b_2 are bias vectors. The intermediate dimension d_{ff} is typically expanded relative to the residual width d, with a standard heuristic of $d_{ff} \approx 4d$. For example, in a model with $d = 512$, d_{ff} is often 2048. In terms of signal processing, this structure is equivalent to two convolutional layers with kernel size 1, often referred to as 1×1 convolutions.

This expansion-compression architecture suggests an interpretation based on basis functions. The first projection W_1 maps the input vector

x into a higher-dimensional space $\mathbb{R}^{d_{ff}}$. We can view the rows of W_1 as a bank of filters or "keys" against which the input is matched. The activation function σ then acts as a selection mechanism, suppressing responses that do not match the filter criteria (in the case of ReLU) or gating them continuously (in the case of GELU or Swish). The second projection W_2 synthesizes the output by linearly combining the activated features. Thus, the FFN acts as a key-value associative memory where W_1 contains the patterns to detect and W_2 contains the content to write back to the residual stream upon detection.

To implement this efficiently, frameworks process the entire matrix X via broadcasted matrix multiplications. The absence of token mixing allows the operation to be parallelized trivially across the T dimension.

```
import torch.nn as nn

class PositionWiseFFN(nn.Module):
    def __init__(self, d_model, d_ff, activation=nn.GELU()):
        super().__init__()
        # W_1: Projects from d_model to d_ff
        self.w1 = nn.Linear(d_model, d_ff)
        # W_2: Projects back from d_ff to d_model
        self.w2 = nn.Linear(d_ff, d_model)
        self.activation = activation

    def forward(self, x):
        # x shape: [Batch, Time, d_model]
        # Operations apply to the last dimension (d_model)
        return self.w2(self.activation(self.w1(x)))
```

While the architectural definition is straightforward, the signal propagation properties of the FFN are critical for the trainability of deep models. We analyze these properties through the lens of the Jacobian matrix. For a single token x, the Jacobian $J_f(x) = \frac{\partial f}{\partial x} \in \mathbb{R}^{d \times d}$ governs how small perturbations in the input propagate to the output. Applying the chain rule to the standard FFN equation yields:

$$J_f(x) = W_2 \, \mathrm{Diag}\big(\sigma'(W_1 x + b_1)\big) \, W_1,$$

where σ' is the first derivative of the activation function applied element-wise.

This Jacobian reveals the input-dependent nature of the FFN's local linearity. The term $\text{Diag}(\sigma'(W_1x+b_1))$ acts as a diagonal gating matrix that modifies the effective rank and singular values of the transformation based on the current activation state.

- *Sparsity and Rank Reduction (ReLU):* If σ is ReLU, $\sigma'(z) \in \{0,1\}$. The diagonal matrix becomes a binary mask indicating which hidden neurons are active. The Jacobian is then a low-rank product $W_2^{(S)}W_1^{(S)}$, where the superscript (S) denotes the submatrices corresponding to active neurons. This explicitly reduces the rank of the gradient signal, filtering out noise in inactive directions but potentially causing "dead neurons" if the sparsity is too high.
- *Curvature and Smoothness (GELU/Swish):* If σ is a smooth approximation like GELU, $\sigma'(z)$ varies continuously. The diagonal matrix scales the singular values of W_2W_1 rather than strictly zeroing them. This ensures that the Jacobian is full-rank almost everywhere, preserving gradient flow even for inputs that fall in the saturation regime of the nonlinearity.

The singular value distribution of $J_f(x)$ is determined by the alignment between the singular vectors of W_1 and W_2 and the entries of the gating matrix. Since $d_{ff} > d$, the product W_2W_1 could theoretically have rank up to d. However, the initialization of weights typically results in a spectrum where the FFN amplifies certain directions (signal) while attenuating others (noise). In deep Transformers, the norm of the Jacobian $\|J_f(x)\|_2$ must be controlled. If $\|J_f(x)\|_2 \gg 1$ consistently, the network suffers from exploding gradients; if $\|J_f(x)\|_2 \ll 1$, it effectively collapses to the identity map via the residual connection (since the total update is $x + f(x)$).

The "channel mixing" perspective also explains the high parameter count of FFNs relative to attention layers. In a standard Transformer block, the parameters are dominated by the FFN (approximately $8d^2$ vs $4d^2$ for attention). This capacity is necessary because the FFN must approximate a general nonlinear function over $\mathbb{R}^d$. Attention merely re-weights existing vectors; FFNs synthesize new ones. The wide intermediate layer d_{ff} provides the necessary width for the universal approximation theorem to hold locally, allowing the network to disentangle complex features that are linearly inseparable in the d-dimensional residual space.

Modern Large Language Models (LLMs) often replace the simple linear-activation-linear structure with Gated Linear Units (GLU). These variants, such as GeGLU or SwiGLU, introduce a multiplicative interaction that increases the expressivity of the layer for a given parameter budget. A generic GLU variant modifies the hidden layer computation by introducing a second projection V_1:

$$f_{\text{GLU}}(x) = W_2\big(\sigma(W_1 x) \odot (V_1 x)\big),$$

where $\odot$ denotes the element-wise Hadamard product. Here, $W_1 x$ controls the "gate" and $V_1 x$ provides the "value." If σ is the sigmoid function, this resembles an analog circuit gate. In LLMs, σ is typically SiLU (Swish) or GELU, and the operation is simply a bilinear interaction modulated by the nonlinearity.

The Jacobian of a GLU-based FFN is significantly more complex due to the product term:

$$J_{\text{GLU}}(x) = W_2\big[\text{Diag}\big(\sigma'(W_1 x) \odot V_1 x\big) W_1 + \text{Diag}\big(\sigma(W_1 x)\big) V_1\big].$$

This derivative contains two parallel paths: one driven by the change in the gate (W_1) and one driven by the change in the value (V_1). This dual pathway allows the gradient to flow through the value path even if the gate is saturated (provided the gate is open), or through the gate path to adjust the gating magnitude. The multiplicative nature implies that the function is effectively quadratic in x (before the final projection), allowing the FFN to model higher-order feature interactions more efficiently than a standard ReLU network.

To maintain comparable computational cost (FLOPs) and parameter counts when moving from a standard FFN to a GLU variant, the intermediate dimension d_{ff} is typically reduced by a factor of $\frac{2}{3}$. This is because GLU requires two input projections (W_1 and V_1) instead of one. Even with reduced width, GLU variants consistently outperform standard FFNs in perplexity metrics, suggesting that the structural bias of multiplicative gating aligns better with the underlying distribution of natural language features.

```
class SwiGLU(nn.Module):
    def __init__(self, d_model, d_ff):
        super().__init__()
        # Combined projection for efficiency: maps d -> 2*d_ff
        # Splits into gate (W1) and value (V1)
        self.w1_v1 = nn.Linear(d_model, 2 * d_ff, bias=False)
        self.w2 = nn.Linear(d_ff, d_model, bias=False)

    def forward(self, x):
        # Project and split
        gate_value, content_value = self.w1_v1(x).chunk(2, dim=-1)
        # Swish Gate: x * sigmoid(x) is SiLU
        hidden = nn.functional.silu(gate_value) * content_value
        return self.w2(hidden)
```

We must also address the bias terms. Original formulations included biases b_1, b_2. Many modern large-scale implementations (e.g., LLaMA, PaLM) omit biases in the FFN linear layers to improve training stability. Removing the bias constrains the FFN such that $f(0) = 0$ (assuming $\sigma(0) = 0$, which holds for ReLU, GELU, Swish). This enforces a stronger "residual" prior: if the input is zero (or close to it), the update is zero, and the residual stream remains unchanged. With biases, the FFN can shift the manifold even for null inputs, which can contribute to variance drift in deep networks.

The FFN acts as a local signal amplifier or dampener. In the context of the residual stream equation $x_{l+1} = x_l + f(x_l)$, the eigenvalues of the update $I + J_f(x_l)$ determine stability. If the real parts of the eigenvalues of $J_f(x_l)$ are largely negative, the residual connection acts as a stabilizing feedback loop. If they are positive, it amplifies signals. The initialization of W_2 is often scaled by $1/\sqrt{L}$ or similar factors to ensure that at the start of training, $f(x)$ is small and $x_{l+1} \approx x_l$, preserving the identity pathway discussed in subsequent sections on residual dynamics.

In summary, the Position-Wise FFN complements the global mixing of self-attention with a local, highly nonlinear feature transformation. It expands the representation into a high-dimensional basis, filters or gates these features using nonlinear activation functions, and projects the result back to the model dimension. Whether using a simple MLP or a Gated Linear Unit, the FFN provides the necessary capacity for the model to memorize facts and compute static relationships, operating purely in the feature domain d without regard for sequence position T.

4.2. LayerNorm as Per-Token Whitening: Geometry, Gradients, and Parameterization

Layer Normalization (LayerNorm) acts as a geometric regularizer within the Transformer block, distinct from the feature-mixing role of the Feed-Forward Network and the token-mixing role of Attention. While Attention and FFNs perform basis changes and information routing, LayerNorm standardizes the signal statistics at the boundaries of these operations. It is a per-token, per-layer whitening operation that decouples the magnitude of the signal from its direction, enabling deeper networks to propagate gradients stably without the extreme sensitivity to initialization variance seen in unnormalized architectures.

We define LayerNorm formally on a single token vector $x \in \mathbb{R}^d$. Unlike Batch Normalization, which computes statistics across the batch dimension B, LayerNorm computes moments strictly along the feature dimension d. This makes the operation independent of other examples in the batch, a crucial property for processing variable-length sequences and for autoregressive generation where future tokens are unknown.

Given a feature vector x, we first compute the scalar mean $\mu(x)$ and variance $\nu(x)$:

$$\mu(x) = \frac{1}{d}\sum_{i=1}^{d} x_i, \qquad \nu(x) = \frac{1}{d}\sum_{i=1}^{d}(x_i - \mu(x))^2.$$

The normalization step projects x to a standardized form $\hat{x}$:

$$\hat{x} = \frac{x - \mu(x)\mathbf{1}}{\sqrt{\nu(x) + \epsilon}},$$

where $\mathbf{1}$ is the all-ones vector and ϵ is a small constant for numerical stability. Finally, the layer applies a learnable affine transformation using parameters $\gamma, \beta \in \mathbb{R}^d$:

$$\mathrm{LN}(x) = \gamma \odot \hat{x} + \beta.$$

Here, $\odot$ denotes the element-wise (Hadamard) product. When applied to the full residual stream matrix $X \in \mathbb{R}^{T\times d}$, this operation is broadcast row-wise.

Implementation details regarding the epsilon term and parameter initialization can subtly impact numerical convergence. Below is a reference implementation illustrating the exact axes of reduction and element-wise affine application.

```
import torch
import torch.nn as nn

class LayerNormManual(nn.Module):
```

```
    def __init__(self, d_model, eps=1e-5):
        super().__init__()
        self.gamma = nn.Parameter(torch.ones(d_model))
        self.beta = nn.Parameter(torch.zeros(d_model))
        self.eps = eps

    def forward(self, x):
        # x shape: [batch, time, d_model]
        # Mean and var computed over the last dimension (d_model)
        mean = x.mean(dim=-1, keepdim=True)
        var = x.var(dim=-1, unbiased=False, keepdim=True)

        # Normalize
        x_hat = (x - mean) / torch.sqrt(var + self.eps)

        # Element-wise affine transformation
        return self.gamma * x_hat + self.beta
```

Geometric Interpretation: Projection and Invariance

Geometrically, the operation $x \mapsto \hat{x}$ can be decomposed into two orthogonal projections. The subtraction of $\mu(x)\mathbf{1}$ projects x onto the hyperplane orthogonal to the vector $\mathbf{1} = [1, 1, \ldots, 1]^\top$. In this zero-mean subspace, the division by $\sqrt{\nu(x)}$ projects the vector onto a hypersphere of radius $\sqrt{d}$.

This geometric rigidification induces specific invariances in the forward pass.

- *Shift Invariance:* $\mathrm{LN}(x + c\mathbf{1}) = \mathrm{LN}(x)$. Adding a constant scalar to every feature does not change the output. This consumes one degree of freedom, removing the "DC component" of the signal.
- *Scale Invariance:* $\mathrm{LN}(\alpha x) = \mathrm{LN}(x)$ for $\alpha > 0$ (ignoring ϵ). Scaling the input vector globally does not affect the output $\hat{x}$.

These invariances imply that the magnitude of the incoming residual stream $\|x\|$ carries no information through the normalization layer itself; only the relative alignment of features (the direction) is preserved. The learnable parameters γ and β then reintroduce scale and shift capabilities. Without γ and β, the output would be constrained to the hypersphere, potentially limiting the expressivity of subsequent layers (e.g., forcing activations into the linear region of a generic nonlinearity). The parameter γ allows the network to selectively amplify specific feature dimensions, essentially reshaping the hypersphere into a hyperellipsoid, while β allows the representation to shift away from the origin, which is necessary for asymmetric activation functions like ReLU or GELU to function as gates.

A modern variant, RMSNorm, simplifies this geometry by removing the mean-centering operation:

$$\text{RMSNorm}(x) = \frac{x}{\sqrt{\frac{1}{d}\sum x_i^2 + \epsilon}} \odot \gamma.$$

RMSNorm projects inputs onto the sphere centered at the origin without first projecting onto the zero-mean hyperplane. This preserves the scale invariance property—which is the primary driver of training stability—while reducing computational overhead. The success of RMSNorm suggests that the re-centering invariance is less critical for deep Transformers than the re-scaling invariance.

Gradient Dynamics and the Jacobian

The stabilizing effect of LayerNorm is best understood by analyzing its Jacobian matrix $J = \frac{\partial \text{LN}(x)}{\partial x}$. This matrix governs how errors backpropagate through the layer. Let us simplify the analysis by ignoring γ and β momentarily (setting $\gamma = \mathbf{1}, \beta = 0$) and focusing on the normalization mapping $f(x) = \hat{x}$.

The Jacobian $J \in \mathbb{R}^{d \times d}$ derives to:

$$J = \frac{\partial \hat{x}}{\partial x} = \frac{1}{\sigma}\left[I - \frac{1}{d}\mathbf{1}\mathbf{1}^\top - \frac{1}{d}\hat{x}\hat{x}^\top\right],$$

where $\sigma = \sqrt{\nu(x) + \epsilon}$.

This expression reveals three critical mechanical properties of gradient flow through LayerNorm:

- *Gain Adaptation:* The Jacobian is scaled by $\frac{1}{\sigma}$. This acts as an implicit learning rate scheduler. If the input x has large variance (large σ), the gradient is attenuated. If the input has small variance, the gradient is amplified. This negative feedback loop keeps the effective step size in the normalized space consistent, regardless of the magnitude of the weights in previous layers.
- *Mean Removal:* The term $\frac{1}{d}\mathbf{1}\mathbf{1}^\top$ subtracts the mean of the gradient. This ensures that the gradient updates do not shift the inputs along the direction $\mathbf{1}$, preserving the distribution centering.
- *Orthogonality to Input:* The term $\frac{1}{d}\hat{x}\hat{x}^\top$ projects the gradient onto the tangent plane of the sphere defined by $\hat{x}$. This removes the component of the gradient parallel to x. Consequently, backpropagation through LayerNorm never updates the *magnitude* of x, only its *direction*.

This orthogonality effectively decouples the optimization of weight magnitude from the optimization of weight direction in upstream layers. Weights can grow without causing exploding gradients because LayerNorm removes the magnitude component from the gradient signal and scales the remaining directional component inversely to that magnitude.

When we reintroduce γ, the effective Jacobian becomes $J_{\text{full}} = \text{diag}(\gamma)J$. If γ is learned to be small, it can throttle gradient flow through specific channels, but the underlying spherical projection geometry remains the dominant factor in signal propagation.

Signal Propagation: Pre-Norm vs. Post-Norm

The placement of LayerNorm relative to the residual connection fundamentally alters the dynamical system defined by the Transformer depth. There are two primary architectural patterns: Post-Norm and Pre-Norm.

Post-Norm Formulation: The original Transformer placed LayerNorm after the residual addition. The recurrence for a block with sublayer F (Attention or FFN) is:

$$x_{l+1} = \text{LN}(x_l + F(x_l)).$$

In this regime, the output of every block is normalized. While this ensures that the input to layer $l + 1$ has unit variance, it complicates gradient propagation. The gradient $\frac{\partial \mathcal{L}}{\partial x_l}$ must pass through the Jacobian of the LayerNorm at every step. As derived above, this Jacobian scales gradients by $1/\sigma$. During initialization, if the variance of the residual branch $F(x_l)$ is non-negligible, the sum $x_l + F(x_l)$ may grow in variance, causing σ to increase and the gradients to vanish exponentially with depth. This necessitates a "warmup" stage in training, where the learning rate starts small to allow the network to align gradients before full-scale optimization begins.

Pre-Norm Formulation: Modern Large Language Models (including GPT-3, PaLM, and LLaMA) predominantly use the Pre-Norm formulation. The recurrence is:

$$x_{l+1} = x_l + F(\text{LN}(x_l)).$$

Here, the normalization is applied only to the input of the sublayer F, while the residual path x_l bypasses the normalization. Expanding this recurrence over L layers yields:

$$x_L = x_0 + \sum_{l=0}^{L-1} F_l(\text{LN}(x_l)).$$

This structure creates a direct "gradient highway" (an identity path) from the final layer back to the input. The Jacobian of the residual block is $I + \frac{\partial F}{\partial x}$. Unlike the Post-Norm case, there is no LayerNorm Jacobian enveloping the sum.

The Pre-Norm architecture changes the signal scale dynamics. Since $x_{l+1} = x_l + \text{update}$, the variance of the residual stream tends to grow linearly with depth (assuming uncorrelated updates). We can approximate $\text{Var}(x_l) \approx l \cdot \text{Var}(\text{update})$. Since the input to the next sublayer is $\text{LN}(x_l)$, the sublayer sees a unit-variance input regardless of the magnitude of x_l. However, the *effective* update size relative to the residual stream diminishes. If x_l has magnitude $\sqrt{l}$, and the update $F(\text{LN}(x_l))$ has magnitude roughly constant (order 1), the relative change to the state is $\frac{1}{\sqrt{l}}$. This acts as an automatic depth-dependent learning rate scaling, stabilizing the training of very deep networks without requiring complex warmup schedules.

However, Pre-Norm is not without trade-offs. Because the final output x_L can have very large variance, a final LayerNorm is required before the vocabulary projection (logits) to bring the representation back to a stable scale for the softmax operation.

Parameterization and Representational Limits

The affine parameters γ and β account for a significant portion of the layer's parameter count ($2d$ parameters per layer), but their role is often misunderstood. They do not merely "undo" normalization.

Consider the interaction with the following ReLU or GELU activation in the FFN. These functions are not scale-invariant. For a standard ReLU, $\text{ReLU}(x) = \max(0, x)$, the shift β determines the "gate threshold." If β shifts the distribution of $\hat{x}$ to be entirely positive, the ReLU becomes a linear identity function. If β shifts it to be entirely negative, the ReLU blocks all signals. Thus, β in LayerNorm (and the bias in the subsequent linear layer) controls the *sparsity* of the activation patterns.

Similarly, γ controls the *saturation* of activations. For a sigmoid-like gate (as in SwiGLU or GEGLU), a large γ pushes values into the saturating regimes (0 or 1), creating binary-like discrete decisions. A small γ keeps values in the linear regime of the sigmoid, facilitating smooth gradient flow.

If one were to remove γ and β (fixing them to 1 and 0), the network would rely entirely on the weight matrices W of the sublayers to perform these scaling and shifting operations. While theoretically possible, this forces the weights W to perform double duty: encoding features and managing signal statistics. Explicit γ, β parameters decouple these concerns, likely improving optimization conditioning.

LayerNorm functions as a differential geometry controller. It enforces a local spherical topology on the feature space, protecting the network from the runaway dynamics of repeated matrix multiplications. Its Jacobian provides an autoregulatory mechanism that damps gradients for large signals and amplifies them for small signals. The choice of Pre-Norm placement leverages this mechanism to protect the sublayers, while the residual stream itself remains an unconstrained, accumulating memory buffer. This separation of *stable computation* (inside the block) and *accumulating memory* (the residual stream) is central to the Transformer's ability to scale to hundreds of layers.

4.3. Residual Pathways as Dynamical Systems: Depth, Lipschitzness, and Gradient Highways

The structural backbone of the Transformer architecture is the residual stream, a continuous vector space that persists across the depth of the model. Unlike traditional feed-forward networks, where each layer acts as a complete transformation of the state space—mapping x_ℓ to a distinct $x_{\ell+1}$—the Transformer architecture models layer outputs as additive updates to a running state. Formally, for a residual block containing a sublayer function F_ℓ (such as self-attention or a feed-forward network), the update rule is given by:

$$x_{\ell+1} = x_\ell + F_\ell(x_\ell).$$

In the context of pre-norm architectures, which dominate modern Large Language Model (LLM) design, the function F_ℓ absorbs the normalization and the core operation, effectively becoming $F_\ell(x) = \text{Core}(\text{LN}(x))$. This formulation reframes the deep network not as a composite function $f_L \circ \cdots \circ f_1(x_0)$, but as a discrete-time dynamical system governing the evolution of token representations. The variable x_ℓ represents the state of a particle (the token) at time step ℓ, and F_ℓ represents the velocity field driving its trajectory through the semantic vector space.

This perspective is essential for understanding how Transformers scale to hundreds of layers. The additive structure creates a "gradient highway" that permits error signals to propagate almost unimpeded from the final loss to the input embeddings, bypassing the vanishing gradient problem that plagued earlier deep architectures like Recurrent Neural Networks (RNNs) and deep Convolutional Neural Networks (CNNs) without skip connections.

Dimensional Compatibility and Isotropism The algebraic requirement of the residual update $x + F(x)$ is that the input and output of F must share the exact same dimension d. This forces the Transformer architecture to be isotropic: the model width d remains constant throughout the depth of the network. While internal operations within F—such as the expansion to d_{ff} in the feed-forward network or the splitting into h heads of dimension d_h in attention—may project the signal into different bases, the final linear projection in every sublayer (commonly denoted as W_O in attention and W_2 in FFNs) must strictly restore the dimension to d.

This constraint creates a unified state space. A vector at layer 1 and a vector at layer 100 live in the same $\mathbb{R}^d$. This compatibility enables the identity map to serve as a meaningful reference frame. In non-residual networks, the basis at layer ℓ is often unrelated to the basis at layer $\ell + 1$ (up to permutation and rotation). In residual networks, $x_{\ell+1}$ is explicitly a modification of x_ℓ, implying that x_ℓ serves as a first-order approximation of $x_{\ell+1}$. The sublayer F_ℓ learns only the *residual*—the difference needed to refine the representation.

Jacobian Dynamics and the Gradient Highway The stability of signal propagation is best analyzed through the Jacobian of the layer update. Let J_ℓ denote the Jacobian of $x_{\ell+1}$ with respect to x_ℓ:

$$J_\ell = \frac{\partial x_{\ell+1}}{\partial x_\ell} = I + \frac{\partial F_\ell}{\partial x_\ell}.$$

During backpropagation, the gradient of the loss $\mathcal{L}$ with respect to the input embeddings x_0 is computed via the chain rule, involving the product of these Jacobians across all L layers:

$$\frac{\partial \mathcal{L}}{\partial x_0} = \left(\prod_{\ell=0}^{L-1} J_\ell \right)^T \frac{\partial \mathcal{L}}{\partial x_L} = \left(\prod_{\ell=0}^{L-1} \left(I + \frac{\partial F_\ell}{\partial x_\ell} \right) \right)^T \frac{\partial \mathcal{L}}{\partial x_L}.$$

In a standard non-residual network, J_ℓ is simply a weight matrix (or a product of weights and derivatives of nonlinearities). If the singular values of these matrices are consistently smaller than 1, the product decays exponentially toward zero (vanishing gradients). If they are larger than 1, the product explodes.

In a residual network, the term $I + \frac{\partial F_\ell}{\partial x_\ell}$ fundamentally alters the spectral properties of propagation. Provided that the update F_ℓ is initialized such that its Jacobian $\frac{\partial F_\ell}{\partial x_\ell}$ has a small spectral norm, the total Jacobian J_ℓ will have singular values clustered around 1. Specifically, if the singular values of $\frac{\partial F_\ell}{\partial x_\ell}$ are bounded by a small ϵ, the singular values of J_ℓ lie in the interval $[1 - \epsilon, 1 + \epsilon]$.

This ensures that the gradient norm is approximately preserved, even across extreme depths. Expanding the product for the gradient at layer ℓ:

$$\frac{\partial \mathcal{L}}{\partial x_\ell} = \frac{\partial \mathcal{L}}{\partial x_{\ell+1}} + \left(\frac{\partial F_\ell}{\partial x_\ell} \right)^T \frac{\partial \mathcal{L}}{\partial x_{\ell+1}}.$$

The first term, $\frac{\partial \mathcal{L}}{\partial x_{\ell+1}}$, represents the gradient flowing directly through the skip connection—the "highway." The second term is the gradient contribution flowing through the sublayer. Even if the sublayer gradients vanish (e.g., due to saturation in the FFN activation or attention collapse), the highway term guarantees that information from the loss function reaches the lower layers. This property decouples the trainability of the network from its depth.

The Discrete-Time ODE Interpretation The residual update formula resembles the Forward Euler discretization of a continuous-time ordinary differential equation (ODE). If we introduce a virtual "time" parameter t and associate layer indices with discrete time steps $t_\ell = \ell \cdot \Delta t$, the residual equation can be rewritten as:

$$\frac{x(t + \Delta t) - x(t)}{\Delta t} = f(x(t), t),$$

where $F_\ell(x) = \Delta t \cdot f(x, t)$. In standard Transformers, Δt is implicitly subsumed into the magnitude of the parameters within F_ℓ.

This dynamical systems view provides a powerful intuition for depth. A deep Transformer is not solving a hierarchical feature extraction problem in the sense of a CNN (where spatial resolution reduces); rather, it is integrating a flow that transforms the geometry of the data manifold. The "time" t represents the complexity of the reasoning process. Shallow layers perform initial processing (lexical and syntactic associations), while deeper layers perform more complex integration (semantic reasoning and context resolution), evolving the particle x along a trajectory defined by the model parameters.

Numerical stability in ODE solvers requires that the step size Δt be small relative to the stiffness of the vector field f. In Transformer training, this corresponds to the requirement that the update $F_\ell(x_\ell)$ should not be too large relative to x_ℓ. If $F_\ell(x_\ell)$ is large, the discrete update may overshoot, leading to erratic trajectories and unstable training. This connects directly to initialization strategies and the role of LayerNorm.

To visualize the spectral implications of residual connections, we can simulate the singular value distribution of the Jacobian for a deep stack. A standard multiplicative network rapidly becomes ill-conditioned, whereas a residual network maintains a spectrum centered at 1.

```
import numpy as np
```

```
def singular_value_evolution(depth=100, dim=256, residual=True):
    # Initialize a random input vector
    # We track the Jacobian accumulation J_total
    J_total = np.eye(dim)

    # Scale factor for stable initialization
    scale = 1.0 / np.sqrt(depth)

    for _ in range(depth):
        # Random Jacobian for a sublayer dF/dx
        # Using random orthogonal matrices for cleaner spectral
     properties
        # in this illustrative example
        H = np.random.randn(dim, dim)
        U, _, Vt = np.linalg.svd(H)
        # Jacobian of sublayer F is scaled
        dF = U @ Vt * scale

        if residual:
            # J_local = I + dF
            J_local = np.eye(dim) + dF
        else:
            # J_local = dF (Standard Feed-Forward)
            # We assume dF is close to orthogonal to be fair
            J_local = U @ Vt

        J_total = J_local @ J_total

    # Return singular values of the total depth Jacobian
    return np.linalg.svd(J_total, compute_uv=False)

# This code is for illustration of the concept.
# Execution would show Residual SVs near 1.0,
# Non-residual SVs diverging or collapsing.
```

Lipschitz Continuity and Sublayer Boundedness For the dynamical system to be robust, the mapping $x \mapsto x + F(x)$ must be well-conditioned. A key concept here is the Lipschitz constant of the sublayer F, denoted K_F. F is K_F-Lipschitz if $\|F(x) - F(y)\| \leq K_F \|x - y\|$ for all x, y.

The Jacobian spectral norm is bounded by the local Lipschitz constant. If $K_F < 1$, the map $G(x) = x + F(x)$ is invertible (by the Banach

Fixed Point Theorem under certain conditions, or simply observing the spectral radius), and the dynamics are contractive or expansive in a controlled manner. However, if K_F is very large, the system becomes chaotic.

Self-attention is not globally Lipschitz because the softmax operation can become arbitrarily sharp as the inputs grow in magnitude, leading to effectively infinite gradients (or zero gradients in saturation). Similarly, the FFN with unbounded activations (like ReLU or GELU) is not globally bounded in output magnitude, although its local derivative is bounded.

This is where Layer Normalization (LayerNorm) becomes critical. As discussed in the previous section, LayerNorm projects the input x onto a manifold of constant variance (roughly a hyper-sphere). By enforcing normalization before the sublayer (Pre-Norm), we ensure that the input to the attention or FFN logic has a bounded norm of $\sqrt{d}$. Consequently, the effective Lipschitz constant of the sublayer is restricted to the local behavior of F on this manifold.

However, the interaction between residual addition and normalization creates a competition for signal dominance. In the Pre-Norm formulation:

$$x_{\ell+1} = x_\ell + F_\ell\big(\mathrm{LN}(x_\ell)\big).$$

As depth ℓ increases, the magnitude of x_ℓ tends to grow. Assuming the updates $F_\ell(\cdot)$ are uncorrelated with x_ℓ and have variance σ^2, the variance of the residual stream grows linearly with depth: $\mathrm{Var}(x_L) \approx L\sigma^2$. Since LayerNorm removes the scale, the effective contribution of the ℓ-th layer to the final output scales as $1/\sqrt{L}$.

This phenomenon implies that in very deep Transformers, later layers contribute incrementally less to the global direction of the vector.

The network naturally settles into a regime where early layers establish the broad semantic space, and later layers perform fine-grained adjustment. This aligns with the *unrolled estimation* view of ResNets: the network iteratively refines its estimate of the target representation.

Initialization and the "Zero-Init" Trick Understanding the residual stream as a dynamical system informs initialization strategies. If we initialize the weights of F_ℓ such that $F_\ell(x) \approx 0$ at the start of training, the network behaves like the identity function $x_L \approx x_0$. This provides a pristine gradient highway, allowing the optimizer to gradually introduce complexity.

A common technique in training deep Transformers (such as GPT-2 and its successors) is to scale the weights of the output projection matrices (W_O in attention, W_2 in FFN) by a factor of $1/\sqrt{2L}$ or similar. This explicitly minimizes the initial impact of the residual branch. In terms of the Jacobian $J = I + \nabla F$, this scaling ensures that $\|\nabla F\| \ll 1$, forcing the singular values of J to be extremely close to 1.

Without this scaling, the variance of the forward pass grows rapidly, and the Jacobian singular values diverge. While LayerNorm mitigates the forward signal explosion, it cannot fix the backward gradient explosion/vanishing caused by a poorly conditioned Jacobian product. The combination of Pre-Norm placement (preserving the identity path) and careful initialization scaling (preserving the identity spectrum) is the cornerstone of stability for decoders with tens or hundreds of layers.

Comparison of Pre-Norm and Post-Norm Dynamics The distinction between Pre-Norm and Post-Norm architectures is essentially a distinction between two different dynamical systems.

- **Post-Norm:** $x_{\ell+1} = \text{LN}(x_\ell + F_\ell(x_\ell))$. Here, the Jacobian is $J_\ell =$

$J_{\text{LN}} \cdot (I + \nabla F_\ell)$. The presence of J_{LN} means we are multiplying by a projection matrix at every step. Since normalization removes information (magnitude), the product of many J_{LN} matrices can rank-collapse or distort the gradient signal. This necessitates a "warm-up" stage in training to carefully align the weights before the gradients become reliable.

- **Pre-Norm:** $x_{\ell+1} = x_\ell + F_\ell(\text{LN}(x_\ell))$. Here, the Jacobian is $J_\ell = I + \nabla F_\ell \cdot J_{\text{LN}}$. Critically, the term I is outside the multiplication. The gradient flow is a sum of the identity path and the branch path. Even if the branch path is distorted by J_{LN}, the identity path remains robust. This structural advantage allows Pre-Norm networks to train without warm-up (or with less sensitivity to it) and to scale to significantly greater depths.

The trade-off is that Pre-Norm networks may have a "representation collapse" issue where the output of the final layer x_L has very large magnitude, making the final LayerNorm (applied before the softmax) carry a heavy burden. However, for signal propagation during training, the benefits of the unimpeded gradient highway provided by the Pre-Norm residual connection outweigh these representational concerns.

By modeling the residual stream as a discretized flow, we transform the heuristics of "skip connections" into a rigorous framework of stability. The Transformer block is not just a feature extractor; it is a time-step in a learned evolution process, carefully engineered to satisfy the conditions of a stable dynamical system. This stability is what allows the simple algebraic operation of addition to support the emergence of complex intelligence in models with billions of parameters.

4.4. Decoder-Only Transformer Block Algebra: End-to-End Dataflow and Interfaces

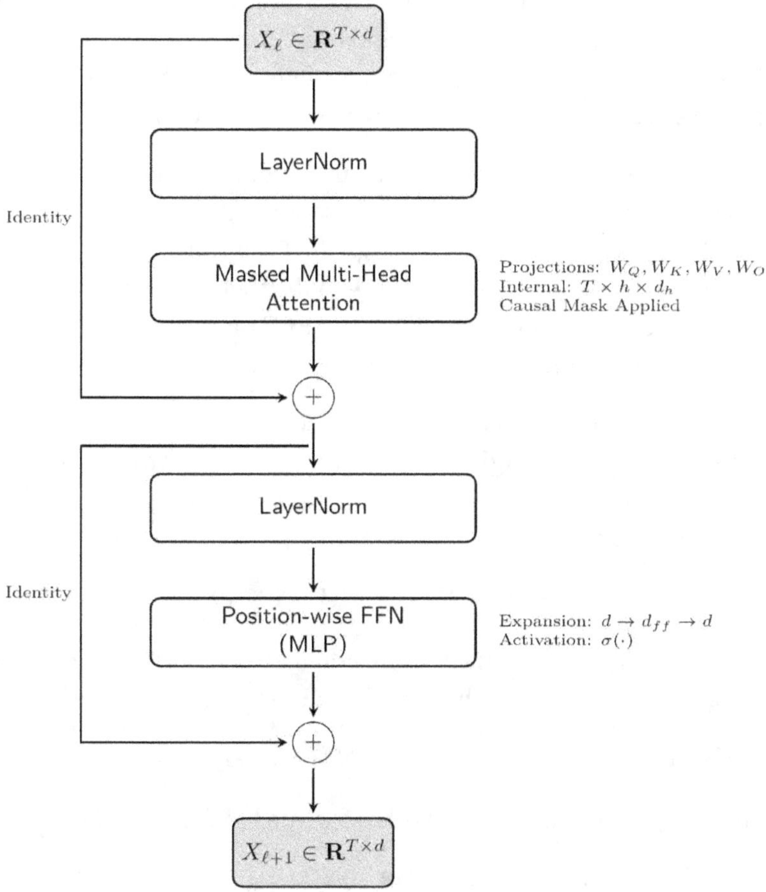

The decoder-only Transformer block is the fundamental atomic unit of modern large language models. While previous discussions have isolated the mechanics of attention, feed-forward networks (FFNs), and normalization, practical implementation requires integrating these components into a single, cohesive operator. This operator, denoted as a function $\mathcal{B} : \mathbb{R}^{T\times d} \to \mathbb{R}^{T\times d}$, transforms a sequence of hidden states while strictly preserving the tensor geometry required for deep stacking. The block architecture defines not only the computational graph but also the optimization landscape; the placement of normalization layers and residual connections determines whether gradients can flow effectively through dozens or hundreds of layers.

We focus here on the "Pre-Norm" formulation, which has become the de facto standard for GPT-style models due to its superior stability in deep networks compared to the original "Post-Norm" design. In this configuration, the block consists of two distinct sublayers-token mixing (self-attention) and channel mixing (FFN)-each preceded by Layer Normalization and followed by a residual addition.

Let $X_\ell \in \mathbb{R}^{T\times d}$ represent the input residual stream to layer ℓ, where T is the sequence length and d is the model width. The processing within a single decoder block proceeds in two stages. First, the attention sublayer processes the normalized stream:

$$Y_\ell = X_\ell + \text{MaskedMHA}\left(\text{LN}_1(X_\ell)\right)$$

Here, Y_ℓ acts as an intermediate residual stream. This intermediate state is then processed by the feed-forward sublayer:

$$X_{\ell+1} = Y_\ell + \text{FFN}\left(\text{LN}_2(Y_\ell)\right)$$

Combining these, the full recurrence for one block is

$$\begin{aligned} X_{\ell+1} = X_\ell &+ \text{MaskedMHA}\left(\text{LN}_1(X_\ell)\right) \\ &+ \text{FFN}\Big(\text{LN}_2(X_\ell + \text{MaskedMHA}(\text{LN}_1(X_\ell)))\Big). \end{aligned}$$

Although the expanded equation appears complex, the algebraic structure reveals a critical property: the output is simply the input plus two distortion terms. If we denote the attention residual branch as f_{attn}

and the FFN residual branch as f_{ffn}, the block updates the stream as $X_{\ell+1} = X_\ell + f_{\text{attn}}(X_\ell) + f_{\text{ffn}}(X_\ell + f_{\text{attn}}(X_\ell))$. This structure ensures that the identity path-the "highway" for gradient propagation-remains uninterrupted.

The first sublayer is the Masked Multi-Head Attention (MHA). This operator is responsible for mixing information across the time dimension T. The interface requires three inputs: Queries (Q), Keys (K), and Values (V). In a decoder-only architecture, there is no cross-attention to an external encoder; therefore, Q, K, and V are all projections of the same input stream. Specifically, let $\tilde{X} = \text{LN}_1(X_\ell)$. We compute:

$$Q = \tilde{X}W_Q, \quad K = \tilde{X}W_K, \quad V = \tilde{X}W_V$$

where $W_Q, W_K, W_V \in \mathbb{R}^{d \times d}$. These projections are logically partitioned into h heads, each of dimension $d_h = d/h$. The attention mechanism computes the scaled dot-product interaction between tokens:

$$A = \text{softmax}\left(\frac{QK^\top}{\sqrt{d_h}} + M\right)V$$

The matrix $M \in \mathbb{R}^{T \times T}$ is the causal mask, which is crucial for the autoregressive property. $M_{ij} = -\infty$ if $j > i$ (preventing positions from attending to the future) and 0 otherwise. This mask is applied additively to the logits prior to the softmax. The resulting matrix A has shape $T \times d$ (after concatenating heads). Finally, an output projection $W_O \in \mathbb{R}^{d \times d}$ mixes the head outputs back into the residual basis:

$$\text{MaskedMHA}(\tilde{X}) = AW_O$$

Dropout is typically applied to this result before it is added to the residual stream X_ℓ. The causality constraint imposed by M ensures that the t-th row of Y_ℓ depends only on rows 1 through t of X_ℓ.

The second sublayer is the Position-wise Feed-Forward Network (FFN), often implemented as a Multi-Layer Perceptron (MLP). This operator mixes information across the feature dimension d but operates independently on each token. It serves to approximate the nonlinear functions governing the evolution of the state vectors. Let $\tilde{Y} = \mathrm{LN}_2(Y_\ell)$. The canonical FFN structure is:

$$\mathrm{FFN}(\tilde{Y}) = \sigma(\tilde{Y} W_{up}) W_{down}$$

where $W_{up} \in \mathbb{R}^{d \times d_{ff}}$ expands the width (typically $d_{ff} \approx 4d$), σ is a nonlinearity such as GELU, and $W_{down} \in \mathbb{R}^{d_{ff} \times d}$ projects back to the residual width. More recent variants, such as SwiGLU, modify this internal structure by adding a gating branch, but the block-level interface remains identical: a map from $\mathbb{R}^{T \times d} \to \mathbb{R}^{T \times d}$ that does not mix information across T.

The integrity of the residual stream is paramount. The tensors X_ℓ, Y_ℓ, and $X_{\ell+1}$ must all possess identical shapes ($T \times d$). This invariant allows blocks to be stacked arbitrarily deep without reshaping. It also implies that the "capacity" of the hidden state is fixed; the model cannot allocate more dimensions to complex tokens than to simple ones. Instead, the model must encode all necessary context-syntactic, semantic, and working memory-within the fixed vector size d. The Pre-Norm placement creates a clean signal path. Because the normalization occurs inside the residual branches ($f(\mathrm{LN}(x))$), the main path $x \to x + \dots$ is never normalized. This prevents the signal magnitude from decaying or exploding purely due to structural reasons, decoupling the optimization of deep layers from the scale of the input embeddings.

We can formalize the implementation of this algebraic block in code. The following listing demonstrates a standard Pre-Norm decoder block, highlighting the modular interfaces and the handling of the causal mask.

```
import torch
import torch.nn as nn
import torch.nn.functional as F

class DecoderBlock(nn.Module):
```

```
    def __init__(self, d_model, n_head, d_ff, dropout=0.1):
        super().__init__()
        # Normalization layers
        self.ln1 = nn.LayerNorm(d_model)
        self.ln2 = nn.LayerNorm(d_model)

        # Sublayer 1: Masked Self-Attention
        self.attn = nn.MultiheadAttention(
            embed_dim=d_model,
            num_heads=n_head,
            dropout=dropout,
            batch_first=True
        )

        # Sublayer 2: Feed-Forward Network
        self.ffn = nn.Sequential(
            nn.Linear(d_model, d_ff),
            nn.GELU(),
            nn.Linear(d_ff, d_model),
            nn.Dropout(dropout)
        )

        # Dropout for residual connections
        self.dropout1 = nn.Dropout(dropout)
        self.dropout2 = nn.Dropout(dropout)

    def forward(self, x, causal_mask=None):
        """
        Args:
            x: Input tensor of shape (Batch, SeqLen, d_model)
            causal_mask: Boolean mask ensuring autoregressive
 property
        Returns:
            Output tensor of shape (Batch, SeqLen, d_model)
        """
        # --- Sublayer 1: Attention ---
        # 1. Normalize input (Pre-Norm)
        x_norm = self.ln1(x)

        # 2. Self-Attention (Q=K=V=x_norm)
        # attn_output shape: (Batch, SeqLen, d_model)
        attn_output, _ = self.attn(
            query=x_norm,
            key=x_norm,
            value=x_norm,
            attn_mask=causal_mask,
            need_weights=False
        )
```

```
        # 3. Residual connection
        x = x + self.dropout1(attn_output)

        # --- Sublayer 2: FFN ---
        # 1. Normalize intermediate (Pre-Norm)
        y_norm = self.ln2(x)

        # 2. Feed-Forward
        ffn_output = self.ffn(y_norm)

        # 3. Residual connection
        x = x + self.dropout2(ffn_output)

        return x
```

This code explicitly maps the algebraic operations to software primitives. Note that `MultiheadAttention` in standard libraries encapsulates the projection matrices W_Q, W_K, W_V, W_O. The `forward` pass clearly exhibits the residual ladder structure: `x = x + branch(norm(x))`. The `causal_mask` is passed into the attention layer, enforcing the $j \leq i$ constraint on the attention matrix.

In a full decoder-only model, these blocks are stacked serially, $X_0 \to X_1 \to \cdots \to X_L$. The input X_0 is formed by summing token embeddings and position embeddings. The final output X_L is typically normalized one last time ($\text{LN}_f(X_L)$) before being projected to the vocabulary size to produce logits. This final normalization is necessary in Pre-Norm architectures because the residual stream X_L has accumulated variance from L additive updates and may have a large magnitude and arbitrary mean, which would destabilize the softmax operation if not standardized.

The computational cost of the block is dominated by dense matrix multiplications. For a sequence of length T and dimension d, the FFN consumes $O(Td^2)$ operations. The attention mechanism consumes $O(Td^2)$ for projections and output, plus $O(T^2d)$ for the attention score computation and aggregation. As T grows, the $O(T^2)$ term in attention becomes the bottleneck, limiting the context window size. However, for the feature dimension d, the block behaves as a densely connected processor where every neuron in layer ℓ has a path to every neuron in layer $\ell + 1$ (via projections and FFN mixing), facilitating rapid information diffusion across the width of the model.

Strict adherence to these interfaces allows the block to be treated as a black box during high-level architectural design. Whether the internal FFN uses ReLU or Swish, or whether the attention uses absolute or rotary position embeddings, the block adheres to the contract: it consumes a length-T sequence of vectors and produces a causally updated length-T sequence of vectors. This modularity is the key to the scalability of Transformer-based LLMs, enabling the training of models with hundreds of billions of parameters by simply replicating this single algebraic unit.

Chapter 5

From Latent States to Probabilities

A Transformer's internal states are not yet probabilities-they are coordinates in a learned feature space. This chapter shows how those coordinates are turned into a calibrated categorical distribution over tokens via a linear readout and softmax, and how temperature reshapes that distribution in mathematically predictable ways. The key message is that "decoding" begins long before sampling: the geometry and numerics of logits determine what probability mass is even available to your inference algorithm (covered in Chapter 7).

5.1. Linear Vocabulary Readout: The Output Projection as a Token Classifier

The final stage of the Transformer architecture, before probability normalization, is the transformation of latent representations into vocabulary-sized scores. Throughout the deep layers of the network, the model maintains a hidden state $h_t \in \mathbb{R}^d$ for each token position t. This vector represents the model's contextual understanding of the sequence up to that point. To generate the next token, this dense, continuous representation must be mapped onto a discrete set of symbols-the vocabulary $\mathcal{V}$. This mapping is performed by the output projection layer, often referred to as the "unembedding" layer or the "LM head."

Mathematically, the readout is an affine transformation mapping the hidden state space $\mathbb{R}^d$ to the logit space $\mathbb{R}^{|V|}$. Let $|V|$ denote the size of the vocabulary, typically ranging from 32,000 to over 100,000 in modern large language models. The projection is parameterized by a weight matrix $W_{\text{out}} \in \mathbb{R}^{|V| \times d}$ and a bias vector $b_{\text{out}} \in \mathbb{R}^{|V|}$. For a single time step t, the logits $z_t \in \mathbb{R}^{|V|}$ are computed as

$$z_t = W_{\text{out}} h_t + b_{\text{out}}.$$

Here, we adhere to the convention where vectors are column vectors, and the matrix multiplication transforms the dimension from d to $|V|$. In implementation frameworks like PyTorch or JAX, where data is typically stored in row-major formats (batch, sequence, features), the operation is often expressed using the transpose W_{out}^T:

$$z_t^T = h_t^T W_{\text{out}}^T + b_{\text{out}}^T.$$

This linear layer acts as the bridge between the semantic space of the model and the symbolic space of the output. The resulting vector z_t contains the unnormalized log-probabilities, or logits, for every possible token in the vocabulary.

The structural simplicity of this operation belies its geometric significance. Each row of the matrix W_{out} corresponds to a specific token in the vocabulary. Let $w_i \in \mathbb{R}^d$ denote the i-th row of W_{out}, corresponding to the i-th token $v_i \in \mathcal{V}$. The logit for this token, $z_{t,i}$, is the dot product of the token's weight vector and the current hidden state, plus the bias:

$$z_{t,i} = \langle w_i, h_t \rangle + b_i.$$

This formulation reveals that the output layer functions as a dense associative memory or a multiclass linear classifier. The vector w_i acts as a "prototype" or "embedding" for token v_i in the output space. The model predicts token v_i with high probability if the current context vector h_t is highly similar to the prototype w_i (in terms of the inner product) relative to all other prototypes w_j.

Tensor Shapes and Batch Processing

In training and inference, we rarely process a single vector h_t. Instead, we operate on batches of sequences. Let B be the batch size and T be the sequence length. The hidden states form a tensor $H \in \mathbb{R}^{B\times T\times d}$. The output projection is applied pointwise to every position in the batch and sequence.

The broadcasting rules of linear algebra libraries allow us to express this transformation succinctly. The weight matrix W_{out} is broadcast across the batch and time dimensions. The resulting logit tensor $Z \in \mathbb{R}^{B\times T\times |V|}$ is computed via a tensor contraction. In Einstein summation notation, this is:

$$Z_{bti} = \sum_{k=1}^{d} H_{btk}(W_{\text{out}})_{ik} + (b_{\text{out}})_i.$$

In standard matrix notation, reshaping H to a 2D matrix of shape $(B \cdot T) \times d$ allows the use of optimized matrix-matrix multiplication (GEMM) routines:

```
# H: [batch_size, seq_len, hidden_dim]
# W_out: [vocab_size, hidden_dim]
# b_out: [vocab_size]

# Flatten batch and time dimensions for matrix multiplication
```

```
H_flat = H.view(-1, hidden_dim)

# Z_flat = H_flat @ W_out.T + b_out
Z_flat = torch.matmul(H_flat, W_out.t()) + b_out

# Reshape back to [batch_size, seq_len, vocab_size]
Z = Z_flat.view(batch_size, seq_len, vocab_size)
```

This transformation dominates the parameter count in small-width models due to the factor $|V| \times d$, but it is computationally straightforward—a large projection. The memory bandwidth required to read W_{out} often constitutes a bottleneck during autoregressive decoding, where $T = 1$ and the arithmetic intensity (ratio of FLOPs to bytes accessed) is low.

Rank Constraints and the Softmax Bottleneck

A critical theoretical constraint arises from the dimensionality mismatch between the hidden state d and the vocabulary size $|V|$. In almost all standard configurations, $d \ll |V|$. For example, a model might have $d = 4096$ and $|V| = 32,000$. Since the logits z_t are generated by a linear map from $\mathbb{R}^d$, the set of all possible logit vectors $\{z_t \mid h_t \in \mathbb{R}^d\}$ forms an affine subspace of dimension at most d within the much larger space $\mathbb{R}^{|V|}$.

This rank limitation, often termed the "softmax bottleneck," implies that the model cannot express arbitrary log-probability distributions over the vocabulary. The matrix of logits for any set of contexts $Z_{\text{matrix}} \in \mathbb{R}^{N \times |V|}$ (where N is the number of contexts) has rank at most d. Consequently, the log-probabilities of tokens are linearly dependent.

Consider the matrix of log-probabilities for the vocabulary. If we treat the log-probabilities as a matrix where rows are contexts and columns are tokens, this matrix can be factored into HW_{out}^T. This factorization enforces a structural rigidity: if the model assigns high probability to "cat" and "dog" in context A, and high probability to "cat" in context B, the linear dependencies constrain the probability of "dog" in context B. The model cannot independently manipulate the score of every token; it must move them in correlated groups defined by the principal directions of W_{out}.

While this might seem restrictive, it acts as a powerful inductive bias. Natural language is structured; tokens that are semantically similar (e.g., "excellent" and "superb") should have correlated probabilities across different contexts. The low-rank structure of the output projection enforces exactly this type of generalization. However, in cases where the true conditional distribution requires high-rank interaction (e.g., highly specific, uncorrelated outcomes), this bottleneck can limit performance. Increasing d (the model width) relaxes this constraint, directly increasing the rank of the representable logit space.

Geometric Interpretation: Prototypes and Angular Margins

The interpretation of $z_{t,i} = \langle w_i, h_t \rangle + b_i$ invites a geometric analysis. If we ignore the bias term and assume normalized vectors, the logit depends on the cosine similarity between the hidden state h_t and the token embedding w_i:

$$z_{t,i} \propto \cos(\theta_{t,i}) = \frac{w_i \cdot h_t}{||w_i|| \, ||h_t||}.$$

Under this view, the output projection partitions the hidden space $\mathbb{R}^d$ into regions. For any given region, one token v_i has the maximum projection score. The boundaries between these regions-where the model is indifferent between token i and token j-are hyperplanes defined by:

$$\langle w_i, h_t \rangle + b_i = \langle w_j, h_t \rangle + b_j \implies \langle w_i - w_j, h_t \rangle + (b_i - b_j) = 0.$$

These decision boundaries form a Voronoi-like tessellation on the surface of the hypersphere (if we consider h_t magnitude fixed, usually by Layer Normalization). The "cell" for token i is the polyhedral cone in $\mathbb{R}^d$ where w_i yields the highest score.

This geometry explains why token embeddings w_i must be distributed effectively throughout the space. If two tokens v_i and v_j have vectors $w_i \approx w_j$, they will always have similar logits, and their probabilities will rise and fall together. This is desirable for synonyms. Conversely, for antonyms or unrelated words that appear in distinct contexts, the vectors w_i and w_j must be nearly orthogonal or opposed to allow the model to distinguish them confidently.

The magnitude of the vectors $\|w_i\|$ also plays a role. A token with a very large norm $\|w_i\|$ will tend to produce extreme logits (highly positive or highly negative) more easily than a token with a small norm. This can function as a learned prior frequency: common tokens might acquire larger norms or biases to ensure they have higher base probabilities when the directional alignment is ambiguous.

Weight Tying: Coupling Input and Output Representations

In many Transformer implementations, the output projection matrix W_{out} is not a separate set of parameters but is shared with the input embedding matrix $E \in \mathbb{R}^{|V| \times d}$. This technique, known as *weight tying*, sets $W_{\text{out}} = E$.

The rationale for weight tying is both practical and theoretical. Practically, it significantly reduces the number of parameters. For a vocabulary of 50,000 and hidden dimension 1,024, the embedding matrix contains roughly 51 million parameters. Duplicating this for the output layer would double the memory footprint of the vocabulary-dependent components.

Theoretically, weight tying enforces a strong semantic constraint: the representation used to *read* a token should be the same as the representation used to *predict* it. If the vector e_{king} encodes the semantic concept of "king" at the input, then the hidden state h_t that predicts "king" should be close to e_{king}.

Algebraically, under weight tying, the logit for token i becomes:

$$z_{t,i} = \langle e_i, h_t \rangle + b_i.$$

(Note: Often the input embedding layer does not have a bias, but the output layer requires one to model unigram frequencies, so a bias b_{out} is learned separately even if W_{out} is tied.)

This coupling implies that the input and output spaces are isometric. However, the roles of e_i are distinct in the two contexts. At the input, e_i is the initial state of the residual stream. At the output, e_i is the target direction. While beneficial for regularization and efficiency, recent research suggests this constraint is not strictly necessary for performance and can sometimes be suboptimal, as the "context" vector h_t (which aggregates information from previous tokens) is logically distinct from a static token identity. Nevertheless, weight tying remains a standard default in architectures like GPT-2 and T5.

Identifiability and Invariance

A crucial property of the logits z_t is that they are over-parameterized with respect to the probabilities they define. The probability distribution is obtained via the softmax function (discussed in detail in the following section):

$$p(y_t = i \mid h_t) = \frac{e^{z_{t,i}}}{\sum_j e^{z_{t,j}}}.$$

Observe what happens if we add a scalar constant c to every logit component: $z'_{t,i} = z_{t,i} + c$.

$$\frac{e^{z_{t,i}+c}}{\sum_j e^{z_{t,j}+c}} = \frac{e^{z_{t,i}} e^c}{\sum_j e^{z_{t,j}} e^c} = \frac{e^c e^{z_{t,i}}}{e^c \sum_j e^{z_{t,j}}} = \frac{e^{z_{t,i}}}{\sum_j e^{z_{t,j}}}.$$

The constant e^c cancels out. The distribution is invariant to uniform shifts in the logits. This means the absolute values of the logits are not identifiable from the observed probabilities; only the *differences* between logits matter. Geometrically, this implies that the effective degrees of freedom in the logit vector z_t is $|V| - 1$. The logits project onto the simplex, and any component of z_t parallel to the vector of all ones $\mathbf{1} = [1, 1, \ldots, 1]^T$ is nullified by the normalization.

This invariance has implications for the output projection parameters. We can shift the bias vector b_{out} by any constant scalar without changing the model's predictions. Similarly, we can shift every row of W_{out} by a constant vector $u \in \mathbb{R}^d$ (i.e., $w_i' = w_i + u$). The change in logit i would be $\langle w_i + u, h_t \rangle = \langle w_i, h_t \rangle + \langle u, h_t \rangle$. Since the term $\langle u, h_t \rangle$ is added to *every* logit, it cancels out during softmax. Thus, the center of mass of the output embeddings is irrelevant to the classification decision.

This redundancy is often handled implicitly during training (the optimizer drifts parameters as needed), but it highlights that the "score" $z_{t,i}$ has no absolute meaning-it is purely a relative quantity representing the preference for token i over the ensemble of other tokens.

Finally, while the linear map W_{out} projects the high-dimensional hidden state to the vocabulary, it serves as the final interface where the continuous, latent reasoning of the Transformer collapses into discrete categorical choices. The properties of this matrix-its rank, the geometry of its rows, and its relationship to the input embeddings-define the boundaries of what the model can express.

5.2. Softmax as Log-Partition: Normalization, Probability Ratios, and Numerical Stability

The linear readout produces a vector of logits $z \in \mathbb{R}^{|V|}$, which represents unnormalized scores for each token in the vocabulary. To transform these scores into a valid probability distribution over the vocabulary, we employ the softmax function. While often treated as a heuristic activation function in introductory contexts, softmax is formally the link function for the categorical distribution in exponential family form. It maps the unconstrained Euclidean space of logits to the $(|V| - 1)$-dimensional probability simplex $\Delta^{|V|-1} = \{p \in \mathbb{R}^{|V|} \mid p_i \geq 0, \sum_i p_i = 1\}$.

The transformation is defined component-wise. For a logit vector z, the probability assigned to token i is:

$$p_i = \text{softmax}(z)_i = \frac{\exp(z_i)}{\sum_{j=1}^{|V|} \exp(z_j)}.$$

This formulation highlights that the absolute magnitude of a logit z_i is meaningless in isolation; it only carries semantic weight relative to the other logits in the vector. Two fundamental properties arise immediately from this definition: shift invariance and the ratio identity.

Shift Invariance. If we add a scalar constant c to every component of z, the resulting probability distribution remains unchanged. Substituting $z_i + c$ into the definition yields:

$$\frac{\exp(z_i + c)}{\sum_j \exp(z_j + c)} = \frac{\exp(z_i) \cdot \exp(c)}{\exp(c) \sum_j \exp(z_j)} = \frac{\exp(z_i)}{\sum_j \exp(z_j)}.$$

This invariance implies that the softmax function is not injective; it is a many-to-one mapping where an entire line $z + c\mathbf{1}$ maps to a single point on the simplex. Geometrically, logits are identified only up to a translation along the vector $\mathbf{1} = [1, 1, \dots, 1]^\top$. This property is critical for numerical stability, as it allows us to center logits arbitrarily without altering the model's predictions.

The Ratio Identity. The relationship between any two outcomes i and k depends exclusively on the difference between their logits. The ratio of their probabilities is:

$$\frac{p_i}{p_k} = \frac{\exp(z_i)/Z}{\exp(z_k)/Z} = \exp(z_i - z_k),$$

where Z is the normalization constant. Taking the logarithm gives the log-odds (or logit) formulation:

$$\log\left(\frac{p_i}{p_k}\right) = z_i - z_k.$$

This identity establishes that $z_i - z_k$ is exactly the log-likelihood ratio of token i versus token k. If z_i increases by δ while z_k remains fixed, the relative probability mass shifts exponentially in favor of i, regardless of the total mass Z. This linear-to-exponential relationship is why logits are the preferred space for unconstrained optimization: additive updates to weights translate directly to multiplicative updates in probability ratios.

To rigorously analyze the analytic and computational properties of softmax, it is advantageous to define the log-partition function (or cumulant function), denoted $A(z)$:

$$A(z) = \log\left(\sum_{j=1}^{|V|} \exp(z_j)\right).$$

We can then rewrite the log-probability of token i as the difference between its score and the global potential:

$$\log p_i = z_i - A(z).$$

Here, $A(z)$ acts as a soft maximum function, often called LogSumExp. As the magnitude of one logit $z_{\max}$ dominates the sum, $A(z) \approx z_{\max}$. The term $A(z)$ captures the "total energy" of the system, and calculating it accurately is central to both the theoretical derivatives and practical implementation of the layer.

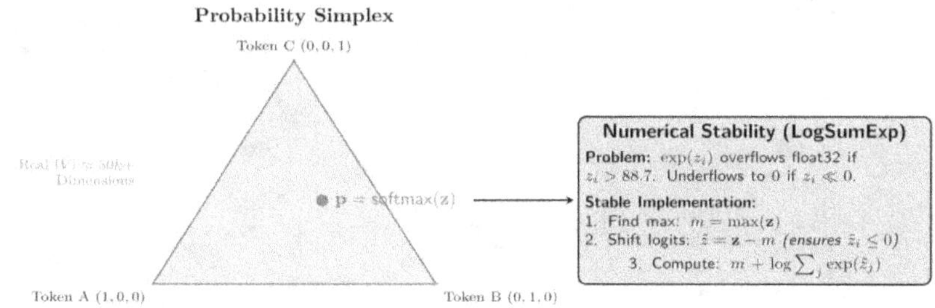

The evaluation of exponentials poses significant numerical risks. In IEEE 754 single-precision floating point (float32), the maximum representable finite value is approximately 3.4×10^{38}, which corresponds to $\exp(88.7)$. If any logit z_i exceeds roughly 89, $\exp(z_i)$ overflows to infinity, causing the entire sum and subsequent division to yield NaN (Not a Number). Conversely, for large negative logits, $\exp(z_i)$ underflows to zero. While underflow is often benign (the probability is effectively zero), overflow is catastrophic.

To prevent this, we utilize the shift invariance property. We define a shifted vector $\tilde{z} = z - m$, where $m = \max_i z_i$. This guarantees that the largest component of $\tilde{z}$ is exactly 0, and all other components are negative. Consequently, $\exp(\tilde{z}_i) \in (0, 1]$ for all i. The computation of the log-partition function becomes:

$$\begin{aligned}
A(z) &= \log\left(\sum_j \exp(z_j)\right) \\
&= \log\left(\sum_j \exp(\tilde{z}_j + m)\right) \\
&= \log\left(\exp(m)\sum_j \exp(\tilde{z}_j)\right) \\
&= m + \log\left(\sum_j \exp(z_j - m)\right).
\end{aligned}$$

This formulation, known as the LogSumExp trick, is standard in all major deep learning frameworks. It ensures that we never explicitly compute the exponential of a large positive number. When computing the probabilities themselves, we similarly compute $p_i = \exp(z_i - m - (A(z) - m))$, keeping all intermediate values within safe bounds.

Implementations must also consider precision during accumulation. Even when using low-precision formats like float16 or bfloat16 for matrix multiplications (the linear projection $Z = HW_{out}^T$), it is strictly required to perform the summation $\sum_j \exp(\tilde{z}_j)$ in float32. Accumulating thousands of small positive probabilities in float16 leads to severe rounding errors due to limited mantissa bits, potentially violating the constraint $\sum p_i = 1$.

```
import torch

def stable_softmax(logits):
    # logits shape: (batch_size, vocab_size)

    # 1. Compute max for numerical stability
    # keepdim=True preserves shape for broadcasting
    m, _ = torch.max(logits, dim=-1, keepdim=True)

    # 2. Shift logits (z - m)
    # The max value becomes 0; others become negative.
```

```
shifted_logits = logits - m

# 3. Compute normalization constant Z_tilde = sum(exp(z - m))
# Accumulate in float32 to prevent precision loss
exp_shifted = torch.exp(shifted_logits)
Z_tilde = torch.sum(exp_shifted, dim=-1, keepdim=True)

# 4. Probabilities: exp(z - m) / Z_tilde
probs = exp_shifted / Z_tilde

# 5. Log-Probabilities: (z - m) - log(Z_tilde)
# This is equivalent to log_softmax
log_probs = shifted_logits - torch.log(Z_tilde)

return probs, log_probs
```

Beyond numerical evaluation, the derivatives of the log-partition function reveal the statistical structure of the distribution. The gradient of $A(z)$ with respect to the logit vector z recovers the probability distribution itself. Differentiating $A(z) = \log \sum_k \exp(z_k)$ with respect to a specific logit z_i:

$$\frac{\partial A}{\partial z_i} = \frac{1}{\sum_k \exp(z_k)} \cdot \frac{\partial}{\partial z_i} \left(\sum_k \exp(z_k) \right) = \frac{\exp(z_i)}{\sum_k \exp(z_k)} = p_i.$$

In vector notation, $\nabla_z A(z) = p$. This confirms a standard property of exponential families: the gradient of the log-partition function yields the expectation of the sufficient statistic (which, for a categorical distribution, is the one-hot indicator vector of the token).

For optimization, we require the Jacobian of the softmax map-how the output probabilities change as logits vary. Let J be the Jacobian matrix where $J_{ik} = \frac{\partial p_i}{\partial z_k}$. Using the quotient rule or the chain rule on $\log p_i = z_i - A(z)$:

$$\frac{\partial \log p_i}{\partial z_k} = \delta_{ik} - \frac{\partial A}{\partial z_k} = \delta_{ik} - p_k.$$

Since $\frac{\partial \log p_i}{\partial z_k} = \frac{1}{p_i} \frac{\partial p_i}{\partial z_k}$, we multiply by p_i to isolate the Jacobian entry:

$$\frac{\partial p_i}{\partial z_k} = p_i(\delta_{ik} - p_k).$$

Thus, the Jacobian is $J = \text{diag}(p) - pp^\top$. This matrix is symmetric and positive semi-definite. It corresponds to the covariance matrix of a multinomial distribution with parameters p (for a single trial). The term $-p_i p_k$ (for $i \neq k$) indicates the competitive nature of the softmax: increasing the logit for token k decreases the probability of token i, proportional to the product of their current probabilities.

This derivative structure is computationally efficient. When backpropagating a loss gradient $\delta_p = \nabla_p \mathcal{L}$ through the softmax layer, we compute the vector-Jacobian product:

$$\delta_z = J^\top \delta_p = (\text{diag}(p) - pp^\top)\delta_p = p \odot \delta_p - p(p^\top \delta_p).$$

This avoids constructing the full $|V| \times |V|$ Jacobian matrix, reducing the operation to $O(|V|)$ complexity. This efficiency is paramount given that vocabulary sizes in modern large language models often exceed 50,000 or 100,000 tokens.

The softmax function, therefore, is not merely a normalization step but a geometric compression. It maps the infinite, unconstrained Euclidean space of $\mathbb{R}^{|V|}$ onto the bounded probability simplex. The shift invariance effectively collapses one degree of freedom, while the exponential function distorts distances: large differences in logits are preserved as near-one probabilities, while small negative differences vanish towards zero. This geometry sets the stage for temperature scaling, where we manipulate the concentration of the distribution by simply scaling the logits prior to this mapping.

5.3. Temperature Scaling and Logit Geometry: Entropy Control Without Changing Ranking

The transformation from continuous latent representations to categorical probabilities is mediated by the softmax function, which maps $\mathbb{R}^{|V|}$

to the probability simplex $\Delta^{|V|-1}$. While the readout projection W_{out} determines the relative scoring of tokens, the "sharpness" of the resulting distribution is often treated as a dynamic property, adjustable at inference time without retraining parameters. This adjustment is formalized through temperature scaling, a one-parameter family of distributions $p^{(\tau)}$ parameterized by a strictly positive scalar τ.

This section analyzes the mathematical structure of temperature scaling, formally defined as:

$$p_i(\tau) = \frac{\exp(z_i/\tau)}{\sum_{j=1}^{|V|} \exp(z_j/\tau)}$$

where $z \in \mathbb{R}^{|V|}$ is the logit vector derived from the affine projection $z = W_{out}h + b$. Although often introduced as a heuristic for controlling generation diversity, temperature scaling has a rigorous geometric and statistical interpretation rooted in the thermodynamics of the Boltzmann distribution. Here, logits correspond to negative energy states, and τ dictates the system's tendency to occupy higher-energy (lower-logit) states. We examine how this scalar control alters entropy, sensitivity, and concentration while maintaining strict invariance over the rank ordering of the vocabulary.

Ranking Invariance and Logit Geometry

The foundational property of temperature scaling is that it decouples the *ordering* of preferences from the *confidence* of those preferences. For any two tokens i and j and any finite positive temperature $\tau > 0$, the ratio of their probabilities is given by:

$$\frac{p_i(\tau)}{p_j(\tau)} = \frac{\exp(z_i/\tau)}{\exp(z_j/\tau)} = \exp\left(\frac{z_i - z_j}{\tau}\right)$$

Taking the logarithm of the ratio recovers the scaled logit difference:

$$\ln p_i(\tau) - \ln p_j(\tau) = \frac{1}{\tau}(z_i - z_j)$$

Since $\tau > 0$ and the exponential function is strictly monotonic, the following equivalences hold:

$$p_i(\tau) > p_j(\tau) \iff \exp(z_i/\tau) > \exp(z_j/\tau) \iff z_i > z_j$$

This implies that the rank order of tokens is invariant under temperature scaling. Specifically, the set of most probable tokens (the argmax set) remains constant for all $\tau \in (0, \infty)$. If a model predicts that "cat" is more likely than "dog" at $\tau = 1$, it retains this preference at $\tau = 0.1$ and $\tau = 100$.

Geometrically, dividing by τ corresponds to a radial scaling of the logit vector z relative to the origin. However, due to the shift invariance of softmax ($\text{softmax}(z) = \text{softmax}(z - c\mathbf{1})$), the effective geometry is defined by the projection of z onto the hyperplane orthogonal to the vector $\mathbf{1}$. Temperature scaling expands or contracts the distance of the logit vector from the "center" of the vocabulary space (the vector of mean logits).

- When $\tau < 1$ (low temperature), the logit differences $z_i - z_j$ are magnified. The distribution moves away from the simplex center toward the vertices.
- When $\tau > 1$ (high temperature), the logit differences are compressed. The distribution contracts toward the uniform distribution at the simplex center.

This linear scaling in logit space translates to exponential scaling in probability ratios. A modest scaling factor of $\tau = 0.5$ squares the probability ratios: if token A was twice as probable as token B, it becomes four times as probable. Conversely, $\tau = 2$ takes the square root of the ratios, reducing a 2:1 advantage to approximately 1.41:1.

Asymptotic Limits and Convergence

The behavior of the distribution at the extremes of the temperature scale—approaching zero and infinity—defines the bounds of the model's expressivity.

The Greedy Limit ($\tau \to 0^+$)

As τ approaches zero, the distribution concentrates entirely on the token(s) with the maximum logit value. To prove this, let $z_{max} = \max_k z_k$ and let $K = \{k \mid z_k = z_{max}\}$ be the set of indices achieving the maximum. We utilize the shift-invariant stable form of the probability:

$$p_i(\tau) = \frac{\exp((z_i - z_{max})/\tau)}{\sum_j \exp((z_j - z_{max})/\tau)}$$

The exponents in the numerator and denominator behave as follows:

- If $i \in K$, then $z_i - z_{max} = 0$, so $\exp(0) = 1$.
- If $i \notin K$, then $z_i - z_{max} < 0$. As $\tau \to 0^+$, the quotient $(z_i - z_{max})/\tau \to -\infty$, and the exponential term vanishes: $\exp(-\infty) \to 0$.

The denominator sum therefore converges to $|K| + 0 = |K|$. Consequently, the probability limit is:

$$\lim_{\tau \to 0^+} p_i(\tau) = \begin{cases} \frac{1}{|K|} & \text{if } i \in K \\ 0 & \text{if } i \notin K \end{cases}$$

In the typical case where the maximum is unique ($|K| = 1$), $p(\tau)$ converges to a one-hot vector (a Dirac delta distribution on the simplex). This corresponds to greedy decoding.

The Uniform Limit ($\tau \to \infty$)

As τ grows large, the scaled logits z_i/τ approach zero. Using the first-order Taylor expansion $e^x \approx 1 + x$ near $x = 0$:

$$p_i(\tau) \approx \frac{1 + z_i/\tau}{\sum_j (1 + z_j/\tau)} = \frac{1 + z_i/\tau}{|V| + (\sum_j z_j)/\tau}$$

Multiplying numerator and denominator by τ (for sufficiently large τ) reveals that the deviations from uniformity are governed by the centered logits:

$$p_i(\tau) \approx \frac{1}{|V|} + \frac{1}{|V|\tau}(z_i - \bar{z})$$

where $\bar{z}$ is the mean logit. As $\tau \to \infty$, the term involving $1/\tau$ vanishes, and $p_i(\tau) \to 1/|V|$. The distribution becomes the maximum entropy uniform distribution, reflecting a state of complete uncertainty where the model's learned preferences are effectively erased.

Entropy Dynamics and Sensitivity

Temperature serves as a direct control mechanism for the Shannon entropy of the output distribution, $H[p(\tau)] = -\sum_i p_i(\tau) \ln p_i(\tau)$. While it is intuitively clear that higher temperatures increase uncertainty, the exact mathematical relationship links the rate of entropy change to the statistical variance of the logits.

Recall that $\ln p_i(\tau) = z_i/\tau - A(z/\tau)$, where A is the log-partition function. Substituting this into the entropy definition:

$$\begin{aligned}
H(\tau) &= -\sum_i p_i(\tau) \left(\frac{z_i}{\tau} - A(z/\tau)\right) \\
&= A(z/\tau) - \frac{1}{\tau} \sum_i p_i(\tau) z_i \\
&= A(z/\tau) - \frac{1}{\tau} \mathbb{E}_p[z]
\end{aligned}$$

Here, $\mathbb{E}_p[z]$ is the expected logit value under the distribution $p(\tau)$. To determine how H changes with τ, we differentiate with respect to τ.

The derivative involves the variance of the logits under the distribution $p(\tau)$, denoted as $\text{Var}_p(z) = \mathbb{E}_p[z^2] - (\mathbb{E}_p[z])^2$. A standard result from the exponential family properties (analogous to heat capacity in thermodynamics) yields:

$$\frac{dH}{d\tau} = \frac{1}{\tau^3}\text{Var}_p(z)$$

Since variance is non-negative and $\tau > 0$, the derivative is non-negative. This provides a rigorous proof that entropy is strictly monotonically increasing with temperature (unless p is already uniform or a point mass, where variance is zero).

The magnitude of this derivative, scaled by τ^3, indicates that entropy is most sensitive to temperature changes when the distribution over logits is "wide"—that is, when there is significant competition between tokens with disparate scores. If the model is already extremely confident (one logit dominates, variance is low) or extremely uncertain (logits are nearly equal, variance is low), temperature adjustments have a diminished effect on the absolute entropy value compared to the intermediate regime.

Gradient Scaling

The sensitivity of the probabilities to the logits themselves is also scaled by temperature. The Jacobian matrix J of the map $z \to p(\tau)$ has entries:

$$\frac{\partial p_i}{\partial z_j} = \frac{1}{\tau}p_i(\tau)(\delta_{ij} - p_j(\tau))$$

The factor $1/\tau$ acts as a global gain. At low temperatures ($\tau \ll 1$), the gradient magnitude can become very large for tokens in the transition region between low and high probability, creating a "hard" decision boundary. Conversely, at high temperatures, gradients are attenuated, and the probability distribution becomes insensitive to perturbations in z. This scaling is crucial when analyzing the stability of decoding: a low temperature makes the output distribution highly sensitive to small noise in the hidden states h_t, potentially leading to flip-flopping decisions if logits for the top candidates are close.

Numerical Stability Implementation

Implementing temperature scaling requires care with floating-point arithmetic, particularly when τ is small. The standard computation involves dividing logits by τ before exponentiation. If τ is close to zero, z_i/τ can exceed the maximum representable float (approx. 88 for float32, leading to $e^{88} \approx 10^{38}$), causing overflow.

The stable softmax approach (subtracting the maximum) must be applied *after* the temperature division or, equivalently, by scaling the subtraction. The correct robust formulation is:

```
def softmax_with_temperature(logits, temperature):
    # Avoid division by zero
    if temperature <= 1e-5:
        # Return one-hot for the argmax
        max_idx = np.argmax(logits)
        p = np.zeros_like(logits)
        p[max_idx] = 1.0
        return p

    # Scale logits first
    scaled_logits = logits / temperature

    # Shift by max(scaled_logits) for stability
    # exp(z/T - max(z)/T) guarantees exponents <= 0
    m = np.max(scaled_logits)
    shifted_logits = scaled_logits - m

    exp_logits = np.exp(shifted_logits)
    return exp_logits / np.sum(exp_logits)
```

Note that simply computing `softmax(logits) / temperature` is mathematically incorrect; the division must occur inside the exponential. Furthermore, the `temperature` parameter must be handled as a hyperparameter of the generation process, distinct from the model weights.

Calibration and ``Energy'' Interpretation

The formal structure of temperature scaling $p_i \propto \exp(z_i/\tau)$ is identical to the Boltzmann distribution in statistical mechanics, where $p_i \propto \exp(-E_i/k_B T)$. In this analogy:

- The logit z_i corresponds to the negative energy $-E_i$ of state i. Higher logit values represent lower energy (more stable/probable) states.
- The parameter τ corresponds to $k_B T$, the thermal energy of the system.

This interpretation clarifies the role of τ as a "noise" level. At zero temperature, the system settles into its ground state (lowest energy / highest logit) deterministically. As thermal energy (τ) is added, the system gains the probability to occupy higher energy states (lower logits).

In the context of probability calibration, temperature scaling is often used as a post-hoc method to align predicted confidence with empirical accuracy. Neural networks, particularly deeply layered transformers, can be uncalibrated—outputting probabilities like 0.99 for events that happen only 80% of the time (overconfidence) or vice versa. By optimizing a single scalar τ on a validation set (minimizing Negative Log Likelihood while keeping W_{out} fixed), one can rescale the "spread" of the logits to match the true error distribution.

Crucially, because τ is a scalar applied globally, it cannot change the relative ranking of errors; it can only adjust the global confidence level. A model that assigns a higher logit to the wrong class than the right class will still make an error at any temperature; temperature scaling simply adjusts how emphatically the model claims that incorrect prediction.

From the perspective of the readout layer, temperature scaling reveals that the magnitude of the logit vector $\|z\|$ is somewhat arbitrary. The training objective (Cross Entropy) encourages $z_{target} \gg z_{other}$, pushing magnitudes to grow until the softmax saturates or weight decay counteracts them. Temperature acts as a runtime renormalization of this magnitude, acknowledging that the "natural" scale of logits learned during training may not be the optimal scale for the decision dynamics required at inference time.

Chapter 6

Optimization: Maximum Likelihood Estimation

A Transformer language model is "just" a probability distribution-so training it should be a matter of fitting that distribution to data. This chapter makes that statement fully precise: we derive the maximum-likelihood objective for token sequences, show how it becomes a sum of per-token cross-entropies under teacher forcing, and connect the resulting loss to perplexity as an interpretable measure of predictive uncertainty.

6.1. From Maximum Likelihood to Token-Level Cross-Entropy (Negative Log-Likelihood)

The foundation of training Large Language Models (LLMs) lies in statistical estimation. Before discussing architectures, optimization dynamics, or parallelization strategies, we must rigorously define the objective function. The goal of pre-training is to construct a model that assigns high probability to linguistically coherent sequences. We formalize this using the framework of Maximum Likelihood Estimation (MLE), which posits that the optimal parameters for a probabilistic model are those that maximize the likelihood of the observed data. In this section, we derive the standard training objective—minimizing the cross-entropy loss over tokens—directly from first principles, demonstrating how the chain rule of probability transforms a sequence-level objective into a sum of local, token-level classification tasks.

The Statistical Estimation Problem

Consider a dataset $\mathcal{D} = \{x^{(1)}, x^{(2)}, \dots, x^{(N)}\}$ consisting of N sequences. Each sequence $x^{(n)}$ is a tuple of discrete tokens $(x_1^{(n)}, x_2^{(n)}, \dots, x_{T_n}^{(n)})$, where each token belongs to a fixed vocabulary V of size $|V|$. The vocabulary V is the discrete sample space defined by the tokenization process (as detailed in Chapter 2). We assume that these sequences are drawn independent and identically distributed (i.i.d.) from an unknown underlying data-generating distribution $p_{\text{data}}(x)$.

Our goal is to learn a parametric model distribution $p_\theta(x)$, parameterized by θ, that approximates $p_{\text{data}}(x)$. Under the MLE framework, we seek the parameters $\hat{\theta}_{\text{MLE}}$ that maximize the joint probability of the observed sequences in the dataset:

$$\hat{\theta}_{\text{MLE}} = \arg\max_\theta \prod_{n=1}^{N} p_\theta(x^{(n)}).$$

Because the product of many probabilities (each strictly between 0 and

1) typically results in values so small they cause floating-point underflow, it is standard practice to maximize the logarithm of the likelihood instead. Since the logarithm is a monotonically increasing function, maximizing the likelihood is equivalent to maximizing the log-likelihood:

$$\mathcal{L}(\theta) = \sum_{n=1}^{N} \log p_\theta(x^{(n)}).$$

This summation represents the total log-probability mass the model assigns to the training corpus.

Autoregressive Factorization via the Chain Rule

The quantity $p_\theta(x^{(n)})$ represents the joint probability of an entire sequence of tokens. Modeling this high-dimensional joint distribution directly is intractable due to the combinatorial explosion of possible sequences. However, the chain rule of probability allows us to decompose any joint distribution into a product of conditional probabilities without loss of generality. For a single sequence $x = (x_1, \ldots, x_T)$, the joint probability factorizes as:

$$p_\theta(x) = p_\theta(x_1) \cdot p_\theta(x_2 \mid x_1) \cdot p_\theta(x_3 \mid x_1, x_2) \cdots p_\theta(x_T \mid x_1, \ldots, x_{T-1}).$$

This factorization is exact and makes no independence assumptions. It expresses the probability of the sequence as the product of the probabilities of each token conditioned on its history. In the context of language modeling, this is often referred to as the autoregressive or causal factorization.

Applying the logarithm to this product converts it into a sum of conditional log-probabilities:

$$\log p_\theta(x) = \sum_{t=1}^{T} \log p_\theta(x_t \mid x_{<t}),$$

where $x_{<t}$ denotes the history $(x_1, \ldots, x_{t-1})$. For the first token ($t = 1$), the history is empty, and the term is simply $\log p_\theta(x_1)$.

This decomposition is the structural key to LLM training. It reframes the problem of "predicting a sequence" into a series of strictly local problems: "given a context, predict the next token." Consequently, the global objective function over the entire dataset becomes a massive sum over all tokens in all sequences:

$$\mathcal{L}(\theta) = \sum_{n=1}^{N} \sum_{t=1}^{T_n} \log p_\theta(x_t^{(n)} \mid x_{<t}^{(n)}).$$

Maximizing this sum is equivalent to training the model to predict x_t accurately given $x_{<t}$ for every position t in the corpus.

From Log-Likelihood to Cross-Entropy

In modern neural network frameworks, we typically minimize a loss function rather than maximizing an objective. Therefore, we define the training loss as the negative log-likelihood (NLL). To understand how this relates to cross-entropy, we must examine the prediction task at a single time step t.

Let the history be $h = x_{<t}$ and the target token be $y = x_t$. The model receives h (processed via embeddings and attention layers) and outputs a vector of unnormalized scores, or logits, $z \in \mathbb{R}^{|V|}$. The model estimates the conditional probability distribution using the softmax function:

$$p_\theta(v \mid h) = \frac{\exp(z_v)}{\sum_{k \in V} \exp(z_k)}, \quad \text{for each } v \in V.$$

The target y is a specific token from the vocabulary. We can represent this target as a probability distribution q over V. Since we are training on observed data where the next token is fixed and known, q is a "one-hot" distribution (also known as a Dirac delta distribution over the discrete set V):

$$q(v) = \begin{cases} 1 & \text{if } v = y \\ 0 & \text{otherwise.} \end{cases}$$

The cross-entropy $H(q, p_\theta)$ between the target distribution q and the model distribution p_θ is defined as:

$$H(q, p_\theta) = -\sum_{v \in V} q(v) \log p_\theta(v \mid h).$$

Because $q(v)$ is zero for all $v \neq y$, the summation collapses to a single term:

$$H(q, p_\theta) = -1 \cdot \log p_\theta(y \mid h) = -\log p_\theta(x_t \mid x_{<t}).$$

Thus, minimizing the negative log-likelihood of the true token is mathematically identical to minimizing the cross-entropy between the empirical one-hot distribution and the model's predicted distribution. This identity holds strictly for hard targets. If we were to use techniques like label smoothing (where q assigns small probability mass to incorrect tokens), the cross-entropy would differ from the pure NLL by a constant entropy term $H(q)$, but for standard pre-training, the equivalence allows us to use the terms interchangeably.

Implementation: LogSoftmax and Numerical Stability

While the mathematical derivation involves computing probabilities via softmax and then taking their logarithm, implementing it this way is numerically unstable. If a logit z_y is very small compared to other logits, $\exp(z_y)$ may underflow to zero, causing $\log(0)$ (infinity) in the loss. Conversely, if z_k is very large, $\exp(z_k)$ may overflow.

To ensure stability, deep learning libraries utilize the `LogSoftmax` operation, which computes $\log p_\theta$ in a single fused step using the Log-Sum-Exp trick:

$$\log p_\theta(y \mid h) = z_y - \log\left(\sum_{k \in V} \exp(z_k)\right) = z_y - \text{LSE}(z).$$

The `CrossEntropyLoss` function in frameworks like PyTorch combines this stable log-softmax computation with the NLL selection. It takes the raw logits z and the target index y as inputs, effectively performing the following operation:

```
def simplified_cross_entropy(logits, target_index):
    # logits: shape (vocab_size,)
    # target_index: integer representing the correct token index

    # Stable LogSumExp
    c = logits.max()
    log_sum_exp = c + torch.log(torch.sum(torch.exp(logits - c)))

    # Log probability of the target class
    log_prob = logits[target_index] - log_sum_exp

    # NLL is negative log probability
    return -log_prob
```

This fused operation ensures that gradients are well-behaved even when probabilities are extremely close to 0 or 1.

Batching, Normalization, and Masking

The derivation so far has produced a sum of losses. In practice, we train using Minibatch Stochastic Gradient Descent (SGD). A minibatch contains B sequences. The total loss for the batch is the sum of the NLLs for every valid token in every sequence. However, simply summing the losses makes the magnitude of the objective function dependent on the batch size and the sequence length, which would require adjusting the learning rate whenever these dimensions change.

To decouple the optimization hyperparameters from the data geometry, we typically average the loss. There are two common ways to normalize:

- *Sequence-level mean:* Calculate the sum of NLLs for each sequence, then average over the batch dimension B. This effectively weights longer sequences more heavily in the gradient update, as they contribute more terms to the loss.

- *Token-level mean:* Calculate the sum of NLLs for all valid tokens in the batch, then divide by the total number of valid tokens. This is the standard approach in LLM training (e.g., GPT-3, LLaMA) because it treats every prediction event as equally important, regardless of the sequence it belongs to.

The concept of "valid tokens" introduces the necessity of masking. Training data is often packed into tensors of fixed shape $(B, T_{\max})$. Sequences shorter than $T_{\max}$ are padded with a special token (e.g., `[PAD]`). These padding tokens are artifacts of the batching process, not part of the data distribution. We must ensure the model is not penalized for failing to predict `[PAD]`, nor should it be encouraged to predict `[PAD]`.

We handle this by applying a loss mask. Let $M_{n,t} \in \{0, 1\}$ be a binary mask where 1 indicates a valid data token and 0 indicates padding (or a prompt region we do not wish to train on). The final scalar loss $J(\theta)$ for a batch is:

$$J(\theta) = \frac{\sum_{n=1}^{B} \sum_{t=1}^{T_{\max}} M_{n,t} \cdot \ell(x_t^{(n)} \mid x_{<t}^{(n)}; \theta)}{\sum_{n=1}^{B} \sum_{t=1}^{T_{\max}} M_{n,t}},$$

where $\ell(\cdot)$ is the per-token cross-entropy loss. In PyTorch's `CrossEntropyLoss`, this masking is typically handled via the `ignore_index` argument. When the target y equals the `ignore_index` (often -100), the loss function returns 0 for that position and excludes it from the denominator of the average.

```
import torch
import torch.nn as nn

# Vocabulary size 1000, batch size 2, sequence length 4
logits = torch.randn(2, 4, 1000, requires_grad=True)

# Targets: indices in [0, 999]. -100 is the ignore_index.
# Sequence 1: [5, 20, 99, -100] (last token is padding)
# Sequence 2: [8, 12, -100, -100] (last two are padding)
targets = torch.tensor([
    [5, 20, 99, -100],
    [8, 12, -100, -100]
])

# Define loss with reduction='mean' (default)
```

```
criterion = nn.CrossEntropyLoss(ignore_index=-100)

# Reshape logits to (Batch * Seq, Vocab) for the loss function
loss = criterion(logits.view(-1, 1000), targets.view(-1))

# The loss is averaged over exactly 5 positions (3 from seq1, 2 from
    seq2).
# The -100 positions do not contribute to the numerator or
    denominator.
```

Relationship to KL Divergence and Entropy

Why is the negative log-likelihood the correct objective? We can view the training process as minimizing the Kullback-Leibler (KL) divergence between the true data distribution p_{data} and the model distribution p_θ. The KL divergence is defined as:

$$\begin{aligned} D_{\text{KL}}(p_{\text{data}} \| p_\theta) &= \mathbb{E}_{x\sim p_{\text{data}}}\left[\log \frac{p_{\text{data}}(x)}{p_\theta(x)}\right] \\ &= \mathbb{E}_{x\sim p_{\text{data}}}[\log p_{\text{data}}(x)] - \mathbb{E}_{x\sim p_{\text{data}}}[\log p_\theta(x)]. \end{aligned}$$

The first term, $\mathbb{E}_{x\sim p_{\text{data}}}[\log p_{\text{data}}(x)]$, is the negative entropy of the data distribution. Since the data distribution is fixed (it is defined by the dataset), this term is a constant with respect to the model parameters θ. The second term is exactly the expected log-likelihood.

Therefore, maximizing the log-likelihood is equivalent to minimizing the KL divergence.

$$\arg\min_\theta D_{\text{KL}}(p_{\text{data}} \| p_\theta) \equiv \arg\min_\theta \left(-\mathbb{E}_{x\sim p_{\text{data}}}[\log p_\theta(x)]\right).$$

By minimizing the cross-entropy loss on our training set, we are effectively pushing the model distribution p_θ toward the empirical data distribution. If the model is expressive enough and the dataset is large enough, p_θ will converge to p_{data}.

The Expected Loss View

The theoretical objective is an expectation over the true data distribution: $\mathbb{E}_{x\sim p_{\text{data}}}[-\log p_\theta(x)]$. In practice, we do not have access to the infinite stream of all possible text p_{data}. Instead, we use the empirical distribution defined by our training corpus $\mathcal{D}$ as a Monte Carlo approximation of the true expectation:

$$\mathbb{E}_{x \sim p_{\text{data}}}[-\log p_\theta(x)] \approx \frac{1}{N} \sum_{n=1}^{N} -\log p_\theta(x^{(n)}).$$

This approximation justifies the use of finite datasets. Furthermore, minibatch SGD can be viewed as an estimation of this expectation using even smaller subsamples. The validity of this approach relies on the Law of Large Numbers; as long as the minibatches are sampled uniformly at random, the gradients computed on the minibatches are unbiased estimators of the gradients of the true expected loss.

This derivation completes the link between the high-level goal—"learn the language"—and the low-level floating-point operations performed on the GPU. We began with the desire to maximize the probability of observed sequences. The chain rule allowed us to break this global probability into local conditional probabilities. The properties of the logarithm allowed us to treat these as a sum of individual losses. Finally, the definition of cross-entropy for one-hot targets provided the specific functional form—Negative Log-Likelihood—that acts as the training signal. The entire pre-training run is, essentially, the minimization of this single scalar value.

6.2. Teacher Forcing as an Indexing Trick: Shifted Targets, Causal Constraints, and Full-Sequence Parallelism

The autoregressive factorization $p_\theta(x) = \prod_{t=1}^{T} p_\theta(x_t \mid x_{<t})$ suggests a sequential process: the distribution for token x_t depends on the history $x_{<t}$. In inference, this dependency is strictly temporal; the model must generate x_1 before it can compute the embedding for x_1 to generate

x_2. However, optimization requires computing gradients over millions of tokens efficiently. Performing a sequential forward pass for every training sequence—waiting for step $t-1$ to complete before computing step t—would drastically underutilize modern hardware accelerators designed for massive matrix parallelism.

We resolve this tension through *teacher forcing*. During training, the model does not condition on its own past predictions (which would require sequential sampling). Instead, it conditions on the ground-truth prefix provided by the dataset. Because the entire ground-truth sequence $x^{(n)}$ is available prior to the forward pass, the conditional distributions for all positions $t \in \{1, \ldots, T\}$ can be computed simultaneously. This transforms the training process from a recurrent loop into a single parallel operation over fixed data tensors.

Implementation relies on a precise alignment of inputs, masks, and targets. We define the input tensor as the sequence of tokens spanning positions 1 to $T-1$ (or 1 to T with padding), and the target tensor as the same sequence shifted by one position.

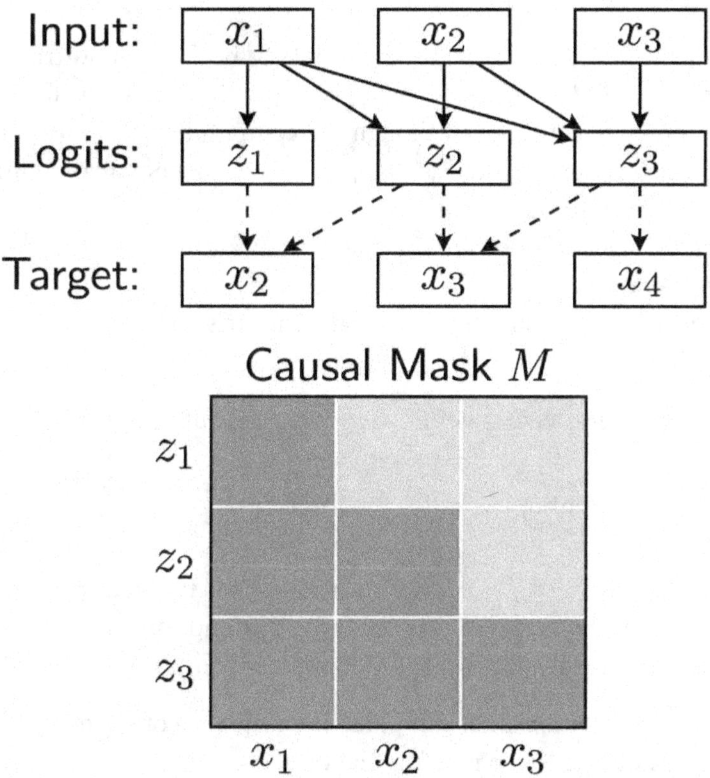

6.2.1. Formalizing the Shifted Objective

Consider a dataset sequence $x = (x_1, x_2, \ldots, x_T)$. The goal is to maximize the likelihood $\prod_{t=1}^{T-1} p_\theta(x_{t+1} \mid x_{\leq t})$. We supply the model with the input tensor $u = (x_1, \ldots, x_{T-1})$. In a standard Transformer architecture, this input passes through an embedding layer and multiple self-attention blocks. Let $H^{(L)}$ denote the output of the final layer L, where $H_t^{(L)} \in \mathbb{R}^d$ corresponds to the representation at position t. This representation is projected via a linear head (the "unembedding" matrix) to produce logits $z_t \in \mathbb{R}^{|V|}$.

Under teacher forcing, we fix the inputs to be the ground truth. Crucially, the validity of this parallel computation relies on the *causal masking constraint*. Even though the computation for z_1 (predicting x_2) and z_{T-1} (predicting x_T) happens in the same matrix multiplication, the mechanism calculating z_t must not have access to any input x_k where $k > t$. If information leaked from future tokens, z_t would no longer represent $p(x_{t+1} \mid x_{\leq t})$ but rather some $p(x_{t+1} \mid x_{\leq t}, x_{>t})$, rendering the likelihood objective invalid and the test-time performance degraded.

In self-attention, this is enforced by the mask matrix M:

$$\text{Attention}(Q, K, V) = \text{softmax}\left(\frac{QK^\top}{\sqrt{d_k}} + M\right) V$$

where $M_{ij} = -\infty$ if $j > i$ and 0 otherwise. This lower-triangular mask ensures that the representation at index i attends only to indices $j \leq i$. Consequently, the logit z_t is a function strictly of $x_1, \ldots, x_t$.

The loss calculation then compares the sequence of logits against the sequence of targets. We define the target sequence y as the input sequence shifted left by one:

$$\begin{aligned} \text{Input } u &= [x_1, x_2, \ldots, x_{T-1}] \\ \text{Target } y &= [x_2, x_3, \ldots, x_T] \end{aligned}$$

The total loss is the sum of the cross-entropies at each position:

$$\mathcal{L}(x) = \sum_{t=1}^{T-1} \text{CrossEntropy}(z_t, y_t)$$

Here, z_t is the prediction made after seeing x_t, and y_t is the actual token that followed x_t.

6.2.2. Implementation: Slicing and Alignment

In practice, training frameworks typically pass a single tensor of shape (B, T) containing the full sequence $[x_1, \ldots, x_T]$ to the model. The model computes logits for all T positions. However, the last logit z_T corresponds to a prediction for x_{T+1}, which does not exist in the truncated sequence. Similarly, the first token x_1 is the target for a hypothetical x_0 (or a Begin-Of-Sentence token), but it acts as input for predicting x_2.

To align these tensors efficiently without copying memory, we use array slicing. The logits are sliced to remove the final position (which has no target), and the targets are sliced to remove the first position (which is not a target for any input provided in the sequence).

The following PyTorch-style snippet demonstrates this pattern. Note how the causal mask is handled internally by the model class (e.g., `GPT2LMHeadModel` or `LlamaForCausalLM`), so the user is responsible primarily for the index alignment during loss computation.

```
def compute_loss(model, input_ids):
    # input_ids shape: (Batch, Length)
    # Forward pass produces logits for all positions
    outputs = model(input_ids)
    logits = outputs.logits  # Shape: (Batch, Length, VocabSize)

    # Shift logits and labels so that:
    # - logits at index t predict token at index t+1
    # - labels at index t are the actual token at index t+1

    # We drop the last logit because we don't have a target for it.
    shift_logits = logits[..., :-1, :].contiguous()

    # We drop the first input token because it is not a target.
    shift_labels = input_ids[..., 1:].contiguous()

    # Flatten batch and sequence dimensions for cross_entropy
    loss = torch.nn.functional.cross_entropy(
        shift_logits.view(-1, shift_logits.size(-1)),
        shift_labels.view(-1)
    )
```

```
    return loss
```

This “shift by one” logic is the standard convention. If the sequence includes a special `[BOS]` token at index 0, then `shift_logits[0]` predicts the token following `[BOS]`, which is the first actual content token.

6.2.3. Handling Padding and Batching

Parallelism introduces a secondary challenge: batching sequences of unequal lengths. We pad shorter sequences with a special token (e.g., `[PAD]`) to match the length of the longest sequence in the batch. While the causal mask handles the autoregressive constraint, we must also ensure that padding tokens do not contribute to the loss. If the model predicts the `[PAD]` token or predicts *from* a `[PAD]` token, these terms should be zeroed out in the objective function.

There are two primary ways to handle masking in the loss function:

- *Label Masking (ignore_index):* The cross-entropy function is configured to ignore a specific target value, usually -100 or the pad token ID. When `shift_labels` contains this value, the gradient for that position is zero.
- *Attention Masking:* We provide an `attention_mask` tensor (1 for real tokens, 0 for padding) to the model. This modifies the self-attention mechanism to prevent real tokens from attending to padding tokens.

While attention masking preserves the correctness of the hidden states (ensuring h_t is not contaminated by future padding), label masking is required for the loss calculation. The model will inevitably produce logits for padding positions, but these logits are meaningless. By setting the targets at these positions to `ignore_index`, we exclude them from the optimization step.

```
def compute_masked_loss(model, input_ids, attention_mask):
    # input_ids: (B, L)
    # attention_mask: (B, L), 1 for text, 0 for pad

    outputs = model(input_ids, attention_mask=attention_mask)
    logits = outputs.logits

    shift_logits = logits[..., :-1, :].contiguous()
    shift_labels = input_ids[..., 1:].contiguous()

    # Check alignment of the attention mask
    # The mask must also be shifted to match the labels
    shift_mask = attention_mask[..., 1:].contiguous()

    # Create the flattened targets, applying ignore_index where mask
     is 0
    # Assuming ignore_index = -100
    flat_labels = shift_labels.view(-1)
    flat_mask = shift_mask.view(-1)

    # Replace padded label IDs with -100
    active_labels = flat_labels.masked_fill(flat_mask == 0, -100)

    loss = torch.nn.functional.cross_entropy(
        shift_logits.view(-1, shift_logits.size(-1)),
        active_labels,
        ignore_index=-100
    )
    return loss
```

This combination of causal constraints (via triangular matrices) and length constraints (via padding masks and label ignoring) allows the autoregressive objective to be fully parallelized. The loop over time t is replaced by a sum over the sequence dimension, enabling the massive throughput characteristic of modern LLM training.

We note a distinct property of this setup known as *exposure bias*. During training, the model is always "corrected" by the teacher (the ground truth data) at every step, regardless of its own prediction. At inference time, however, the model must rely on its own generated history. While this discrepancy theoretically affects long-term generation stability, the efficiency gains of parallel training vastly outweigh the downsides for large-scale pretraining. The strict likelihood objective ensures that minimizing the shifted-target loss is equivalent to minimizing the KL divergence between the data distribution and the model's autoregressive distribution.

6.3. Perplexity as Exponentiated Cross-Entropy: Units, Bases, and Comparability Across Tokenizations

The optimization of autoregressive language models relies fundamentally on the minimization of the negative log-likelihood (NLL) over a training corpus. While gradients and parameter updates are computed directly from the NLL, the magnitude of this loss—measured in nats or bits—can be abstract and difficult to interpret intuitively. Perplexity emerges as the standard intrinsic evaluation metric because it transforms the logarithmic loss back into a linear probability space, offering a concrete interpretation related to the model's uncertainty. However, perplexity is frequently misunderstood as a universal quality metric; it is, in fact, strictly tied to the specific vocabulary and tokenization scheme used by the model. This section formalizes the derivation of perplexity from cross-entropy, establishes the precise relationship between logarithmic bases, and rigorously analyzes the conditions under which perplexity comparisons are valid.

Derivation and the Geometric Mean Interpretation

Perplexity is defined as the exponentiation of the mean negative log-likelihood per token. To be precise, consider a dataset $\mathcal{D}$ consisting of a total of M tokens. In practice, this dataset is often partitioned into sequences, but for the purpose of metric definition, we view it as a stream of tokens $x_1, \dots, x_M$. The autoregressive model assigns a probability $p_\theta(x_t \mid x_{<t})$ to each token given its history. The total negative

log-likelihood of the corpus is

$$\mathcal{L}_{\text{total}}(\mathcal{D}, \theta) = -\sum_{t=1}^{M} \ln p_\theta(x_t \mid x_{<t}).$$

The standard training objective is the arithmetic mean of this sum, commonly referred to as the cross-entropy loss:

$$\bar{\ell} = \frac{1}{M}\mathcal{L}_{\text{total}}(\mathcal{D}, \theta) = -\frac{1}{M}\sum_{t=1}^{M} \ln p_\theta(x_t \mid x_{<t}).$$

Perplexity (PPL) is the exponential of this mean loss:

$$\text{PPL}(\mathcal{D}, \theta) = \exp(\bar{\ell}) = \exp\left(-\frac{1}{M}\sum_{t=1}^{M} \ln p_\theta(x_t \mid x_{<t})\right).$$

Using the properties of logarithms and exponentials, we can rewrite this expression to reveal a fundamental interpretation:

$$\text{PPL}(\mathcal{D}, \theta) = \exp\left(\sum_{t=1}^{M} \ln p_\theta(x_t \mid x_{<t})^{-1/M}\right) = \prod_{t=1}^{M} p_\theta(x_t \mid x_{<t})^{-1/M}.$$

This product form is the *inverse geometric mean* of the conditional probabilities assigned to the observed tokens. If a model assigns a probability of 1.0 to every correct token, the geometric mean is 1, and the perplexity is 1 (the minimum possible value). If a model assigns a low probability to the correct tokens, the inverse geometric mean grows.

The geometric mean is the appropriate aggregate because probabilities are multiplicative. A model that assigns probability 0.5 to two events has a joint probability of 0.25. The arithmetic mean of the probabilities (0.5) would not capture the compounding uncertainty, whereas the geometric mean preserves the relationship to the sequence likelihood.

Interpretation as Effective Branching Factor

The value of perplexity can be interpreted as the *effective branching factor* of the model's predictive distribution. To visualize this, consider an idealized model U_k that, at every time step, possesses zero knowledge about the specific correct token but has narrowed the possibilities down to a set of k plausible candidates. If the model assigns a uniform probability $1/k$ to each of these candidates and 0 to all others, and if the true token is always within this set, then the probability assigned to the observed token is always $1/k$.

The cross-entropy (NLL) for this uniform model is

$$\bar{\ell} = -\ln(1/k) = \ln k,$$

and the perplexity is

$$\text{PPL} = \exp(\ln k) = k.$$

Thus, a perplexity of 20 implies that the model is as confused as if it were choosing uniformly and independently among 20 options at each step. This analogy holds even though the model's actual output distribution is rarely uniform and usually assigns mass to thousands of tokens. The perplexity aggregates high-confidence predictions (where $p \approx 1$) and low-confidence predictions (where $p \ll 1$) into a single scalar representing the average number of equiprobable choices that would result in the same total entropy.

Logarithmic Bases: Nats vs. Bits

The numerical value of the loss $\bar{\ell}$—and consequently the perplexity—depends on the base of the logarithm used. In the context of deep learning frameworks like PyTorch or TensorFlow, the function `log` computes the natural logarithm (base e). The units of $\bar{\ell}$ in this case are *nats*. In information theory, it is common to use base 2, where the units are *bits* (or shannons).

Let $\bar{\ell}_e$ denote the loss in nats and $\bar{\ell}_2$ denote the loss in bits. They are related by the change-of-base formula:

$$\bar{\ell}_2 = \frac{\bar{\ell}_e}{\ln 2} \approx 1.4427 \cdot \bar{\ell}_e.$$

Perplexity is invariant to the choice of base, provided the exponentiation matches the logarithm:

$$\text{PPL} = \exp(\bar{\ell}_e) = 2^{\bar{\ell}_2}.$$

To verify this:

$$2^{\bar{\ell}_2} = 2^{\bar{\ell}_e/\ln 2} = (e^{\ln 2})^{\bar{\ell}_e/\ln 2} = e^{\bar{\ell}_e}.$$

This invariance allows researchers to report perplexity without explicitly specifying the base, unlike cross-entropy, where the unit (nats vs. bits) is critical. However, when converting between reported cross-entropy scores and perplexity, one must ensure the bases align. A common error is computing exp(loss) when the loss was reported in bits, or 2^{loss} when the loss was in nats, leading to drastically incorrect perplexity values.

Metric Hygiene: Aggregation and Masking

Correctly computing perplexity requires rigorous handling of sequence boundaries and padding. The mean NLL $\bar{\ell}$ must be calculated over *valid* tokens only. If a batch of sequences includes padding tokens to align lengths, including these padding positions in the denominator M will artificially lower the loss (since padding is usually trivial to predict, often assigned 1.0 probability deterministically or ignored in the numerator but counted in the denominator).

Consider a batched evaluation. The correct procedure is to sum the negative log-probabilities of all non-padding tokens in the batch and divide by the total count of non-padding tokens.

```
import torch
import torch.nn.functional as F

def compute_perplexity(logits, targets, pad_id):
    """
    logits: (batch_size, seq_len, vocab_size)
    targets: (batch_size, seq_len)
    pad_id: int, identifier for padding token
```

```
    """
    # 1. Compute element-wise cross-entropy (NLL)
    # reduction='none' preserves the loss per token
    loss_per_token = F.cross_entropy(
        logits.view(-1, logits.size(-1)),
        targets.view(-1),
        reduction='none'
    )

    # 2. Create a mask for valid tokens (non-padding)
    # Reshape mask to match flattened loss
    mask = (targets.view(-1) != pad_id).float()

    # 3. Sum valid losses
    total_nll = (loss_per_token * mask).sum()

    # 4. Count valid tokens
    total_tokens = mask.sum()

    # 5. Compute mean NLL and Perplexity
    mean_nll = total_nll / total_tokens
    perplexity = torch.exp(mean_nll)

    return perplexity.item()
```

A critical aggregation error involves computing the perplexity for each sequence independently and then taking the arithmetic mean of those perplexities. Because of the convexity of the exponential function, the arithmetic mean of perplexities is always greater than or equal to the perplexity of the aggregated dataset (Jensen's inequality). The "average perplexity" of a corpus is strictly the exponential of the average NLL, not the average of the exponentials.

To demonstrate the magnitude of this error, consider two tokens with probabilities 0.1 and 0.001.

- Sequence 1: $p(x_1) = 0.1 \implies \text{NLL} \approx 2.30, \ \text{PPL} = 10.$
- Sequence 2: $p(x_2) = 0.001 \implies \text{NLL} \approx 6.91, \ \text{PPL} = 1000.$
- **Incorrect Average PPL**: $(10 + 1000)/2 = 505.$

- **Correct Aggregate PPL**: Mean NLL $= (2.30 + 6.91)/2 \approx 4.605$. $\exp(4.605) \approx 100$.

The incorrect arithmetic mean (505) drastically overestimates the model's perplexity (100) because it allows the worst-case outlier (1000) to dominate linearly rather than logarithmically.

Comparability Across Tokenizations

The most significant limitation of perplexity is that it is a *per-token* metric. Consequently, its value is intrinsically tied to the definition of a "token." This dependency renders raw perplexity scores incomparable across models that use different tokenizers or vocabulary sizes.

Consider the sentence: "The biological data."

- **Word-level tokenizer**: Tokens: ["The", "biological", "data"]. Length $T = 3$.
- **Subword tokenizer (BPE)**: Tokens: ["The", "bio", "logical", "data"]. Length $T = 4$.
- **Character-level tokenizer**: Tokens: ["T", "h", "e", " ", "b", "i", "o", ...]. Length $T = 19$.

Let the total probability of the sequence assigned by the model be $P(S)$. This probability $P(S)$ measures how well the model predicts the string itself, ideally independent of segmentation. The total Negative Log-Likelihood is $\mathcal{L}_{\text{total}} = -\ln P(S)$.

The reported perplexity depends on the denominator M:

$$\text{PPL} = \exp\left(\frac{\mathcal{L}_{\text{total}}}{M}\right).$$

For the word-level model ($M = 3$), the denominator is small, resulting in a larger exponent and a much higher perplexity. For the character-level model ($M = 19$), the denominator is large, reducing the mean NLL and resulting in a much lower perplexity. A character-level model might achieve a perplexity of 3.0, while a word-level model on the same text achieves 100.0. This does not mean the character model is "better"; it simply means the prediction task per step (predicting `h` after `T`) is entropically easier than predicting "biological" after "The". The character model makes many easy predictions, diluting the average surprise.

Formally, perplexity measures the entropy rate per event in the discrete event space defined by the vocabulary. Since the event spaces differ, the rates are incommensurable.

Normalization: Bits per Character and Bits per Byte

To compare models with different tokenizations, we must normalize the total information content by a constant unit inherent to the data, typically a *character* or a *byte*, rather than the variable unit of a *token*.

The standard metric for this comparison is **Bits per Byte (BPB)** or **Bits per Character (BPC)**.

$$\text{BPB} = \frac{\mathcal{L}_{\text{total}} \text{ (in bits)}}{\text{Number of Bytes in original text}}.$$

If the loss is computed in nats, we apply the conversion:

$$\text{BPB} = \frac{\mathcal{L}_{\text{total}} \text{ (in nats)}}{\ln(2) \cdot \text{Number of Bytes}}.$$

This metric is invariant to tokenization. Whether the model predicts the string in 3 steps or 19 steps, if the total assigned probability $P(S)$ is the same, the numerator $\mathcal{L}_{\text{total}}$ is identical. Since the denominator (bytes) is a property of the string S, the BPB score allows for a direct comparison of the density estimation quality of a character-level model versus a subword-level model.

When reading literature, one can approximately convert perplexity to bits-per-token using $\log_2(\text{PPL})$. However, one cannot convert

$\log_2(\text{PPL})$ to BPB without knowing the average number of bytes per token for that specific tokenizer on that specific text. This ratio, $\frac{\text{bytes}}{\text{token}}$, is the compression rate of the tokenizer.

$$\text{BPB} \approx \frac{\log_2(\text{PPL})}{\text{Average Bytes per Token}}.$$

This relationship elucidates why subword models (like BPE or WordPiece) report higher perplexities than character models but are often superior density estimators. Their "average bytes per token" is higher (compressing more characters into a single decision), so they incur more entropy *per step*, but often less entropy *per byte*.

Intrinsic vs. Extrinsic Evaluation

Finally, while perplexity is the mathematically correct objective for maximum likelihood estimation, it is an intrinsic metric. It measures strictly how well the model fits the probability distribution of the test set. It acts as a necessary but not sufficient condition for generation quality. A model with infinite capacity could memorize the training set (low training PPL) but fail on the test set (high test PPL). Conversely, a model might achieve low test perplexity by being very conservative—assigning broad, flat distributions that never miss the target but never commit strongly to a coherent narrative.

In the context of the optimization chapter, perplexity serves as the primary diagnostic for convergence. A failure of the perplexity to decrease indicates a failure of the optimization process (learning rate, gradients) or a capacity issue, independent of the downstream utility of the generated text. It allows the engineer to debug the probabilistic engine in isolation from the decoding algorithms or prompt engineering strategies that will be applied later.

Chapter 7

Inference and Decoding Algorithms

Training defines a distribution; decoding chooses how you traverse it. This chapter shows that "generation quality" is not a mysterious property of the model alone but a mathematically precise consequence of how we repeatedly (i) compute logits, (ii) transform/truncate a categorical distribution, and (iii) sample or select a token that irreversibly changes the future distribution. The central message is that decoding is an algorithmic lens on the same learned probabilities-changing the lens changes the behavior.

7.1. The Autoregressive Decoding Loop as a Stochastic Process on Contexts

Foundational to the study of large language models is the distinction between the probabilistic model learned during training and the procedure used to generate text during inference. While training minimizes a loss function over a fixed dataset, generation is an open-ended process where the model acts as a transition operator within a dynamic system. We formalize this by defining the autoregressive decoding loop not merely as a software implementation, but as a stochastic process defined over the space of possible contexts.

Let V be a finite vocabulary of size $|V|$. We denote a sequence of tokens (a context) as $x_{1:t} = (x_1, x_2, \ldots, x_t) \in V^t$. The core object provided by the language model is the conditional probability distribution $p_\theta(x_{t+1} \mid x_{1:t})$, parameterized by weights θ. This distribution assigns a probability mass to every token $v \in V$ given the history. Through the chain rule of probability, the joint probability of a sequence $x_{1:T}$ is fully specified by the product of these conditional steps:

$$P_\theta(x_{1:T}) = \prod_{t=0}^{T-1} p_\theta(x_{t+1} \mid x_{1:t}).$$

However, this factorization describes the *likelihood* of a sequence, not necessarily the mechanism by which it is produced. In generation, we introduce a decoding policy π that intervenes between the model's raw probabilities and the selection of the next token.

Model Potentials versus Decoding Policies

It is rigorous to view the language model as a function that maps a context $x_{1:t}$ to a set of logits $z_t \in \mathbb{R}^{|V|}$. As detailed in Chapter 5, these logits are converted into a categorical distribution via the softmax function:

$$p_\theta(v \mid x_{1:t}) = \frac{\exp(z_{t,v})}{\sum_{u \in V} \exp(z_{t,u})}.$$

We refer to this as the *model distribution*. It represents the network's learned belief about the next token.

The *decoding distribution*, denoted q_t, is derived from p_θ by a transformation T_π. The policy π dictates how the model's beliefs are operationalized. For example, a greedy policy collapses the distribution to a Dirac delta at the mode, while temperature scaling flattens or sharpens the density. Formally, at each step t:

$$q_t(\cdot) = T_\pi\big(p_\theta(\cdot \mid x_{1:t})\big).$$

The next token X_{t+1} is then sampled from q_t. This distinction is critical: p_θ is a static property of the weights and context, whereas q_t is a dynamic object constructed at runtime. The separation allows us to analyze "model errors" independently from "decoding artifacts" where the policy distorts p_θ excessively, perhaps truncating valid options.

The Transition Kernel

We can now define the generation process as a Markov chain on the state space of all finite strings $\mathcal{S} = \bigcup_{t=0}^{\infty} V^t$. Unlike typical Markov chains with a fixed transition matrix, the transition kernel here is infinite and structured by the string concatenation operation.

Let $s_t = x_{1:t}$ be the state at time t. The transition kernel $K_\pi(s_{t+1} \mid s_t)$ is non-zero only if s_{t+1} is formed by appending a single token v to s_t. Specifically:

$$K_\pi(s' \mid s) = \begin{cases} q_\pi(v \mid s) & \text{if } s' = s \oplus v, \\ 0 & \text{otherwise,} \end{cases}$$

where $\oplus$ denotes concatenation and $q_\pi(v \mid s)$ is the probability assigned to token v by the policy given context s.

The trajectory of the system is the sequence of states $s_0, s_1, s_2, \ldots$. Since s_{t+1} contains s_t as a prefix, the state space forms a tree (a prefix tree or trie), and decoding is a random walk down this tree starting from the root (or a provided prompt). The randomness of the walk is governed entirely by q_t.

The following Python-like pseudocode illustrates this abstraction, highlighting the separation between the logit computation and the policy application.

```
def autoregressive_step(model, policy, context):
    """
    Performs one step of the decoding process.

    Args:
        model: Function mapping context -> logits
        policy: Function mapping logits -> next_token_distribution
        context: Current sequence of token IDs

    Returns:
        next_token: The sampled integer token ID
        next_dist: The probability distribution used for sampling
    """
    # 1. Model creates the raw potentials (logits)
    logits = model(context)

    # 2. Policy transforms logits into the target distribution q_t
    #    (e.g., softmax, top-k truncation, temperature scaling)
    next_dist = policy(logits)

    # 3. Sample from the policy distribution
    next_token = sample(next_dist)

    return next_token
```

Path Dependence and Distribution Shift

A profound consequence of this autoregressive structure is that the sequence of distributions encountered during generation, $\{q_0, q_1, q_2, \dots\}$, is itself a stochastic process. The distribution q_{t+1} depends on x_{t+1}, which was sampled from q_t.

This creates a feedback loop. If the policy samples a low-probability token at step t (a "tail event"), the context $x_{1:t+1}$ moves into a region of the latent space that may be less familiar to the model. Often, this results in higher entropy for q_{t+1} or a shift in mass towards generic or repetitive tokens. We quantify this behavior using the entropy of the decoding distribution:

$$H(q_t) = -\sum_{v \in V} q_t(v) \log q_t(v).$$

Tracing $H(q_t)$ over time reveals the model's fluctuating confidence. A sudden spike in entropy often indicates that the generation has "gone off the rails," entering a context where the model has no strong prediction for the continuation. Conversely, a collapse in entropy suggests the model has entered a rote phrase or a repetitive loop.

The divergence between the model's raw preference p_θ and the policy's choice q_t measures the "intervention strength" of the decoding algorithm. For instance, if π is a top-k sampler, it zeroes out the probability of all tokens outside the top k. The total probability mass removed, $\epsilon_t = \sum_{v \notin \text{top-}k} p_\theta(v \mid x_{1:t})$, represents the cumulative likelihood of the branches pruned by the policy. If ϵ_t is large, the decoder is actively fighting the model's natural inclination, which forces the trajectory onto a path the model considers unlikely.

Operational Constraints: Context Windows and Caching

In practice, the mathematical ideal of a context $x_{1:t}$ growing infinitely is constrained by the model's architecture. Transformer models have a finite context window L. When $t > L$, the condition $p_\theta(\cdot \mid x_{1:t})$ is approximated by $p_\theta(\cdot \mid x_{t-L+1:t})$. The state space of our stochastic process effectively becomes V^L rather than $\bigcup V^t$, and the transition kernel becomes a mapping $V^L \to V^L$ (a sliding window).

This truncation implies that the process is not strictly Markovian on the full history if dependencies extend beyond L steps. However, for the purpose of decoding algorithms, we treat the visible context window as the complete state.

Computational efficiency requires that we do not recompute the embeddings for the entire prefix $x_{1:t}$ at every step. This is achieved via Key-Value (KV) caching. The state is physically represented not just by the token indices, but by the cached attention matrices for all previous layers. Formally, the transition becomes:

$$(z_t, \text{Cache}_{t+1}) = \text{Model}(x_t, \text{Cache}_t).$$

While KV caching is an implementation detail regarding time complexity ($O(1)$ per token vs $O(t)$), it reinforces the view of decoding as a stateful evolution. The "state" is the KV cache, and the "action" is the selection of the next token which updates that cache.

Termination and Absorption

The decoding process must eventually terminate to produce a finite output. We model this by including a special End-of-Sequence (EOS) token in the vocabulary. The state space includes a set of absorbing states: any sequence ending in EOS is terminal. The probability of termination at step t is exactly $q_t(\text{EOS})$.

Failure to terminate-generating infinite loops or rambling text-corresponds to a trajectory that never hits the absorbing set. This often occurs because the probability assigned to EOS, $p_\theta(\text{EOS} \mid x_{1:t})$, drops to near zero in repetitive loops. Decoding algorithms often impose a hard horizon (maximum length T_{max}), effectively forcing a transition to the absorbing state if $t = T_{max}$.

By formalizing decoding as a stochastic process, we move beyond viewing generation as a simple "forward pass." Instead, it is a dynamic interaction between the model's learned manifolds and the policy's constraints, evolving step-by-step through the vast combinatorial space of language.

7.2. Deterministic Decoding via Myopic Optimization: Greedy Argmax and Its Failure Modes

We begin our analysis of decoding strategies with the deterministic baseline: greedy decoding. While often treated as a trivial default, greedy decoding serves as the boundary condition for all stochastic sampling methods. Understanding its behavior, specifically why it fails to generate high-probability sequences despite maximizing probability at every step, provides the theoretical justification for the more complex distributional transformations discussed later in this chapter.

Formally, greedy decoding is defined as the policy π_{greedy} that selects the token with the highest conditional probability mass at each time step t. Given a model parameterized by θ and a context $x_{1:t}$, the next token x_{t+1} is determined by

$$x_{t+1} = \mathbf{argmax}_{v \in V}\, p_\theta(v \mid x_{1:t}).$$

This definition contains a subtlety regarding determinism. The argmax operator returns a set, not a scalar, if multiple tokens share the exact same maximal probability. While rare in trained models using floating-point logits, ties can occur in masked models, quantized environments, or initializations. To render π_{greedy} a strictly deterministic function $f : V^t \to V$, one must impose a canonical tie-breaking rule, such as selecting the token with the lowest integer index in the vocabulary. Without this specification, "greedy decoding" remains ill-defined across different hardware backends that may handle sorting stability differently.

The following implementation enforces canonical determinism by explicitly handling ties, ensuring that the function behaves identically regardless of the underlying accelerator's reduction algorithms.

```
import torch

def canonical_greedy_step(logits: torch.Tensor) -> int:
    """
    Selects the argmax token. Breaks ties by choosing the
    lowest vocabulary index.

    Args:
        logits: Tensor of shape [vocab_size]
    """
    # Identify the maximum value
    max_val, _ = torch.max(logits, dim=-1)

    # Create a boolean mask of all indices matching the max
    # Floating point comparison requires a tolerance epsilon
    is_max = torch.abs(logits - max_val) < 1e-6

    # nonzero() returns indices ordered by value
    candidates = is_max.nonzero(as_tuple=True)[0]

    # Deterministic selection: first candidate (lowest index)
    return candidates[0].item()
```

The Myopic Optimization Trap

The fundamental theoretical limitation of greedy decoding is that it performs coordinate-wise optimization on a sequence objective, which does not guarantee a global optimum. The goal of maximum a posteriori (MAP) decoding is to find the complete sequence $x_{1:T}$ that maximizes the joint probability:

$$\hat{x}_{1:T} = \text{argmax}_{x'_{1:T}} \, p_\theta(x'_{1:T}) = \text{argmax}_{x'_{1:T}} \prod_{t=1}^{T} p_\theta(x'_t \mid x'_{<t}).$$

Greedy decoding approximates this by solving

$$x_t = \text{argmax}_{v \in V} \, p_\theta(v \mid x_{<t}).$$

This is a myopic (shortsighted) approximation. By locking in the locally optimal token at step t, the algorithm constrains the conditioning context for step $t+1$ to a specific subspace of the joint distribution. If the globally optimal sequence requires transitioning through a lower-probability token at step t to access a high-probability region at step $t+1$, greedy decoding will miss it.

We can prove this suboptimality with a minimal counterexample. Consider a vocabulary $V = \{A, B\}$ and a sequence length $T = 2$. Let the model probabilities be defined as follows:

1. *Step 1:* $P(A \mid \text{start}) = 0.6$, $P(B \mid \text{start}) = 0.4$.
2. *Step 2 (given A):* $P(A \mid A) = 0.1$, $P(B \mid A) = 0.9$.
3. *Step 2 (given B):* $P(A \mid B) = 0.99$, $P(B \mid B) = 0.01$.

Under greedy decoding:

- At $t = 1$, the policy compares $P(A) = 0.6$ and $P(B) = 0.4$. It selects A.
- The context is now (A). At $t = 2$, the policy maximizes $P(\cdot \mid A)$, selecting B (0.9).
- The generated sequence is (A, B).
- The joint probability is $P(A, B) = P(A) \times P(B \mid A) = 0.6 \times 0.9 = 0.54$.

Now consider the path starting with B, which was rejected at $t = 1$:

- Select B (suboptimal locally). Context is (B).
- At $t = 2$, maximize $P(\cdot \mid B)$, selecting A (0.99).
- The sequence is (B, A).

- The joint probability is $P(B, A) = P(B) \times P(A \mid B) = 0.4 \times 0.99 = 0.396$.

Wait, in this specific numerical setup, the greedy path (A, B) has probability 0.54, which is higher than (B, A) at 0.396. The greedy assumption holds here. To construct a failure case, we must sharpen the conditional probability in the second branch to overcome the initial deficit. Let us adjust the Step 2 probabilities for the B branch:

Revised Counterexample:

1. *Step 1:* $P(A \mid \text{start}) = 0.51$, $P(B \mid \text{start}) = 0.49$.
2. *Step 2 (given A):* The distribution is uniform entropy. $P(A \mid A) = 0.5$, $P(B \mid A) = 0.5$.
3. *Step 2 (given B):* The distribution is sharp. $P(A \mid B) = 1.0$, $P(B \mid B) = 0.0$.

Under greedy decoding:

- Select A $(0.51 > 0.49)$.
- Select A or B (tie-break). Let us assume A.
- Sequence (A, A). Joint probability: $0.51 \times 0.5 = 0.255$.

Optimal MAP sequence:

- Choose B.
- Choose A (1.0).
- Sequence (B, A). Joint probability: $0.49 \times 1.0 = 0.49$.

Here, $0.49 > 0.255$. The greedy policy fails to identify the mode of the joint distribution because it cannot "look ahead" to see that the slightly less likely token B leads to a deterministic, high-confidence outcome, whereas A leads to uncertainty. This phenomenon is often visualized as a "Garden Path" problem: the parser (or decoder) commits to an interpretation early that turns out to be a dead end or a low-probability valley.

Search Interpretation and Probability Cliffs

Greedy decoding can be interpreted as a Depth-First Search (DFS) on the tree of all possible sequences, with a branching factor of $k = 1$ and a heuristic that strictly follows the edge with the highest weight at each node. Unlike beam search, which maintains a frontier of k hypotheses to smooth over local dips in probability, greedy decoding has no backtracking mechanism.

This structural rigidity makes greedy decoding susceptible to "probability cliffs." A probability cliff occurs when the argmax token x_{t+1} shifts the context $x_{1:t+1}$ into a region of the latent space where the model's predictive distribution $p_\theta(\cdot \mid x_{1:t+1})$ becomes high-entropy (flat) or assigns mass primarily to generic, "safe" tokens.

In natural language generation, this manifests as the well-known "blandness" or "dullness" artifact. Models trained with maximum likelihood estimation (MLE) often assign the highest probability to safe, high-frequency words (e.g., "the", "it", "is", "nice") because these words appear in the widest variety of contexts. While a more specific word might be appropriate for the specific nuance of a prompt, the specific word splits probability mass with its synonyms, whereas the generic word aggregates mass from many potential syntactic continuations. Greedy decoding systematically prefers these generic aggregators. Over long sequences, this compounds, leading to output that is grammatically perfect but semantically empty.

Repetition as a Dynamical Systems Failure

The most severe artifact of deterministic decoding is repetitive looping. We can analyze this using the state-space perspective established in the previous section. Let S be the space of effective contexts. For a Transformer, the effective context is technically the entire history, but in practice, the attention mechanism often focuses on a finite window of recent tokens.

If the decoding process enters a state $x_{1:t}$ such that the generated token x_{t+1} creates a new state $x_{1:t+1}$ that is functionally equivalent to an earlier state $x_{1:t-k}$ (in terms of the induced next-token distribution), the system enters a limit cycle. Because the policy is deterministic, $f(x) \to y$ implies that returning to state x guarantees y is produced again.

If $p_\theta(\cdot \mid x_{t-k:t}) \approx p_\theta(\cdot \mid x_{t:t+k})$ then $x_{t+k+1} = x_{t+1}$.

Once a cycle is established (e.g., "I don't know. I don't know. I don't know."), there is no source of stochasticity to kick the system out of the attractor basin. The probability of the repetition sequence $P(\text{phrase})^N$ may be vanishingly small compared to a diverse sequence, but the greedy decoder never evaluates the sequence probability; it only evaluates the transition probability. If $P(\text{“I”} \mid \text{“...know.”})$ is locally maximal, the loop persists indefinitely until the maximum context length is reached.

Engineering Invariants and Instability

Despite these failure modes, greedy decoding possesses engineering properties that make it a critical baseline.

- *Minimal Entropy / Zero Temperature Limit:* Greedy decoding corresponds to sampling with temperature $\tau \to 0$. As τ approaches zero, the softmax distribution approaches a Kronecker delta (one-hot) distribution at the index of the maximum logit.

- *Reproducibility:* It is the only decoding method that guarantees the exact same output for the same input without managing random number generator (RNG) seeds. This makes it indispensable for regression testing of model pipelines.

- *Instability near Ties:* While deterministic, greedy decoding is highly sensitive to perturbations when the top logits are close. If $z_i \approx z_j$, a change in floating-point precision (e.g., FP32 vs. BF16) or a minor model update can flip the sign of $z_i - z_j$. This flip changes the selected token, which changes the context for all subsequent steps, potentially leading to a completely divergent generation trajectory. This "butterfly effect" implies that greedy outputs from similar models are not necessarily similar, even if the models' distributions are nearly identical in terms of KL divergence.

The limitations of greedy decoding—its myopia, susceptibility to repetition, and tendency toward blandness—necessitate the use of the stochastic sampling methods detailed in the subsequent sections. However, these stochastic methods effectively trade the guarantee of local optimality for the possibility of global coherence, relying on the model's calibrated uncertainty to navigate the garden paths that greedy search ignores.

7.3. Top-k Truncation as Support-Restricted Renormalization of a Categorical Distribution

The autoregressive generation process described previously relies on a decoding policy π to convert the model's conditional distribution

$p_\theta(\cdot \mid x_{1:t})$ into a realized next token. While greedy decoding provides a deterministic baseline by collapsing the distribution to its mode, stochastic sampling aims to preserve the generative diversity of the model. However, sampling directly from the full distribution p_θ carries the risk of selecting low-probability "tail" tokens. In large vocabularies (often $|V| \in [30{,}000, 200{,}000]$), the aggregate mass of tens of thousands of implausible tokens can be non-negligible. While individually unlikely, their collective probability allows standard multinomial sampling to occasionally divert the generation into incoherent or repetitive trajectories.

Top-k sampling addresses this by strictly enforcing zero probability for all but the k most likely tokens. This operation is not merely a heuristic for noise reduction; it is a specific non-linear transformation of the probability simplex that enforces sparsity while preserving the relative ordering of the surviving candidates.

Mathematical Formulation

Let the vocabulary be denoted by V. At time step t, the model produces a categorical distribution $p_t(v) = p_\theta(v \mid x_{1:t})$ over $v \in V$. We define the rank of a token v based on its probability mass, with ties broken deterministically (e.g., by lowest index). Let $\text{rank}_t(v) \in \{1, \ldots, |V|\}$ be this rank, such that $\text{rank}_t(u) < \text{rank}_t(v) \implies p_t(u) \geq p_t(v)$.

The top-k policy identifies a support set $S_k \subset V$ containing the highest-probability tokens:

$$S_k = \{v \in V \mid \text{rank}_t(v) \leq k\}$$

The policy constructs a new distribution $q_t^{(k)}$ by truncating the support to S_k and renormalizing the probability mass. Let Z_k be the cumulative probability mass of the top-k set under the original model distribution:

$$Z_k = \sum_{v \in S_k} p_t(v) = 1 - \sum_{v \notin S_k} p_t(v)$$

The derived distribution is then given by:

$$q_t^{(k)}(v) = \begin{cases} \dfrac{p_t(v)}{Z_k} & \text{if } v \in S_k \\ 0 & \text{if } v \notin S_k \end{cases}$$

This transformation creates a distribution that is absolutely continuous with respect to p_t on the restricted set S_k, but singular with respect to p_t on $V \setminus S_k$.

Preservation of Relative Odds

A critical mathematical invariant of top-k sampling is the preservation of relative likelihoods among the retained tokens. For any two tokens $u, v \in S_k$, the ratio of their probabilities under the policy $q_t^{(k)}$ is identical to their ratio under the model p_t:

$$\frac{q_t^{(k)}(u)}{q_t^{(k)}(v)} = \frac{p_t(u)/Z_k}{p_t(v)/Z_k} = \frac{p_t(u)}{p_t(v)}$$

This property ensures that the truncation process does not distort the model's preference between valid candidates. If the model assigns u twice the probability of v, and both are within the top k, the sampling mechanism will still select u twice as often as v. The policy acts solely as a filter for the tail, leaving the "head" of the distribution structurally intact, merely scaled by the scalar factor $1/Z_k$.

Renormalization Dynamics and Mass Distortion

The scaling factor $1/Z_k$ represents the degree of distortion applied to the retained probabilities. This factor is dynamic; it depends on the shape (sharpness or flatness) of the original distribution p_t, which varies with context.

Consider two distinct regimes:

- **Peaked Distribution (Low Entropy):** If the model is highly confident, the top k tokens may account for nearly all probability mass ($Z_k \approx 1$). In this case, $q_t^{(k)}(v) \approx p_t(v)$ for $v \in S_k$. The truncation is non-invasive, serving mainly to guard against extremely rare anomalies.

- **Flat Distribution (High Entropy):** If the model is uncertain, the mass may be spread thinly across thousands of tokens. Here, Z_k may be significantly less than 1. Consequently, $1/Z_k$ becomes a large multiplier, artificially inflating the probabilities of the top k tokens. For instance, if $Z_k = 0.5$, the probability of every retained token doubles.

This variance in Z_k reveals a limitation of fixed-k strategies: the amount of "valid" probability mass discarded ($1 - Z_k$) fluctuates. A fixed $k = 10$ might be too permissive when the model is certain (admitting 9 absurd tokens if the top 1 holds 99% mass) and too restrictive when the model is uncertain (cutting off valid alternatives if the top 10 only sum to 20% mass).

Algorithmic Implementation via Logits

In practice, neural networks output logits $z \in \mathbb{R}^{|V|}$ rather than probabilities. Top-k truncation is efficiently implemented in the logit space before the softmax operation. Since the softmax function is strictly monotonic, the top k probabilities correspond to the top k logits.

The standard implementation involves sorting or partial sorting (using a selection algorithm like Introselect) to find the k-th largest logit value, $z_{(k)}$. A mask vector $M \in \mathbb{R}^{|V|}$ is constructed:

$$M_v = \begin{cases} 0 & \text{if } z_v \geq z_{(k)} \\ -\infty & \text{if } z_v < z_{(k)} \end{cases}$$

The policy distribution is then obtained by applying softmax to the masked logits:

$$q_t^{(k)} = \text{softmax}(z + M)$$

Adding $-\infty$ ensures that the exponential of the masked logits becomes zero, effectively removing them from the support. The softmax normalization term automatically computes the required Z_k (implicitly) relative to the unnormalized exponents.

```
import torch
import torch.nn.functional as F

def top_k_sampling(logits, k):
    # logits: [batch_size, vocab_size]
    # Find the k-th largest value to determine the cutoff
    top_values, _ = torch.topk(logits, k)
    kth_best = top_values[:, -1].unsqueeze(-1)

    # Create a mask for logits below the cutoff
    # We use -inf to force probability to 0 after softmax
    mask = logits < kth_best
    masked_logits = logits.masked_fill(mask, float('-inf'))

    # Re-normalize over the valid set
    probs = F.softmax(masked_logits, dim=-1)

    # Sample from the support-restricted distribution
    next_token = torch.multinomial(probs, num_samples=1)
    return next_token
```

This implementation avoids explicit renormalization division by leveraging the softmax property $\frac{e^{z_i}}{\sum_{j \in S_k} e^{z_j}} = q_t^{(k)}(i)$.

Entropy Reduction and Calibration

Top-k sampling invariably reduces the entropy of the next-token distribution, denoted as $H(q_t^{(k)}) \leq H(p_t) - \log Z_k$ (approximately, though the exact relationship depends on the tail shape). By zeroing out the tail, the algorithm trades recall (the ability to generate diverse, rare tokens) for precision (high likelihood of coherence).

This support restriction impacts the calibration of the generation. A language model is calibrated if the predicted probability p matches the empirical frequency of the event. Top-k truncation intentionally uncalibrates the output: a token with model probability 0.1 might be sampled with probability 0.2 if $Z_k = 0.5$. This reflects a deliberate design choice in open-ended generation: humans often perceive the high-entropy "true" distribution as containing too many incoherent digressions, preferring the focused subspace defined by S_k.

The boundary conditions of k recover other standard decoding algorithms. As $k \to 1$, S_k shrinks to the singleton set containing the argmax, and $q_t^{(k)}$ degenerates to a Dirac delta function, recovering greedy decoding. As $k \to |V|$, S_k encompasses the entire vocabulary, $Z_k \to 1$, and $q_t^{(k)}$ approaches the pure ancestral sampling distribution p_t. The parameter k thus acts as a control lever for the "randomness budget" permitted in the transition kernel.

The rigidity of the parameter k—remaining constant regardless of the distribution's shape—can lead to text degeneration. When p_t is flat, k may be too small, truncating the distribution mid-plateau and forcing the model to choose from an arbitrarily limited subset, potentially breaking local coherence or grammar. Conversely, when p_t is sharp, a large k may include the very tail elements the method was designed to avoid. This tension motivates adaptive truncation strategies where the size of the support set $|S|$ varies dynamically based on the cumulative probability mass.

7.4. Nucleus (Top-p) Sampling: Minimal Cumulative-Mass Sets and Random Support Size

We formalize nucleus sampling, commonly referred to as top-p sampling, as a distribution-adaptive truncation method that addresses the rigidity of rank-based (top-k) selection. While top-k truncation enforces a fixed cardinality on the candidate set regardless of the model's predictive uncertainty, nucleus sampling enforces a fixed constraint on the cumulative probability mass. This shift from cardinality-based to mass-based control fundamentally alters the stochastic properties of the decoding process, creating a dynamic support set that expands and contracts in response to the entropy of the next-token distribution.

Mathematical Formulation of the Nucleus

Let $P(v) = p_\theta(v \mid x_{1:t})$ denote the conditional probability distribution over the vocabulary V at time step t. To define the nucleus, we first define a ranking of the vocabulary based on probability mass. Let $v_{(1)}, v_{(2)}, \ldots, v_{(|V|)}$ be an ordering of the vocabulary such that

$$P(v_{(1)}) \geq P(v_{(2)}) \geq \cdots \geq P(v_{(|V|)}).$$

This ordering is deterministic; ties in probability are resolved via a fixed precedence rule (e.g., lexicographic order or lowest integer ID) to ensure the decoding policy remains a well-defined function.

We define the cumulative distribution function over the rank statistics as the sequence of partial sums C_m:

$$C_m = \sum_{i=1}^{m} P(v_{(i)}).$$

Given a target probability mass $p \in (0, 1]$, nucleus sampling identifies the smallest subset of tokens-the "nucleus"-that accounts for at least p proportion of the total probability. The cardinality of this set, denoted m^*, is the minimal rank index such that the cumulative mass constraint is satisfied:

$$m^* = \min\{m \in \{1, \ldots, |V|\} \mid C_m \geq p\}.$$

The support set S_p is then defined as the top m^* tokens:

$$S_p = \{v_{(1)}, v_{(2)}, \ldots, v_{(m^*)}\}.$$

The decoding policy π_{nuc} constructs the proposal distribution $q^{(p)}$ by restricting the original model distribution P to S_p and renormalizing.

For any token $v \in V$:

$$q^{(p)}(v) = \begin{cases} \dfrac{P(v)}{Z_p} & \text{if } v \in S_p, \\ 0 & \text{otherwise,} \end{cases}$$

where the normalization constant Z_p is the actual cumulative mass of the selected set:

$$Z_p = \sum_{v \in S_p} P(v) = C_{m^*}.$$

A critical distinction from generic truncation is the boundary condition. The definition implies that $Z_p \geq p$. Because the distribution is discrete, equality $Z_p = p$ rarely holds exactly unless p is carefully chosen to match partial sums of the specific distribution. In most cases, $Z_p > p$, meaning the nucleus policy retains slightly more mass than requested to ensure the boundary token $v_{(m^*)}$ is included. This inclusion is necessary to satisfy the condition $C_m \geq p$; excluding $v_{(m^*)}$ would result in a mass of $C_{m^*-1} < p$.

Dynamic Support Size and Adaptivity

The defining characteristic of nucleus sampling is that the size of the support set, $|S_p| = m^*$, is a random variable dependent on the shape of the distribution P. This contrasts with top-k sampling, where the support size is a constant hyperparameter k.

Consider two limiting regimes of the model's prediction:

- *Low Entropy (High Confidence):* The model assigns most probability mass to a small number of tokens. For example, if $P(v_{(1)}) = 0.95$ and $p = 0.9$, then $C_1 = 0.95 \geq 0.9$. The minimal m^* is 1. The nucleus contains only the single most probable token, and the sampling process effectively degenerates to greedy decoding for this step (or a very sharp Bernoulli if $P(v_{(1)})$ was slightly less than p but combined with $v_{(2)}$ crossed the threshold).

- *High Entropy (Low Confidence):* The distribution is flat, perhaps uniform over a subset of feasible continuations. If the distribution is approximately uniform over N tokens, such that $P(v) \approx 1/N$, then to accumulate mass p, the method must select approximately $p \cdot N$ tokens. If $N = 1000$ and $p = 0.9$, the support size $|S_p|$ grows to roughly 900.

This adaptivity aligns the decoding search space with the model's internal uncertainty. When the model is confident, nucleus sampling restricts the search path to the high-confidence region, preventing the selection of low-probability "tail" tokens that might be reachable simply because the vocabulary is large. When the model is uncertain, nucleus sampling expands the candidate pool, acknowledging that many valid continuations exist and preserving the diversity inherent in the distribution.

In contrast, a fixed top-k policy risks two distinct failure modes. If k is small but the distribution is flat, top-k artificially truncates valid options, forcing the model to choose from an arbitrarily restricted set (over-truncation). If k is large but the distribution is sharp, top-k admits a long tail of effectively zero-probability garbage tokens, which, due to the sheer number of such tokens in large vocabularies, might cumulatively carry enough mass to be sampled occasionally (under-truncation). Nucleus sampling avoids these modes by defining the truncation boundary in probability space rather than index space.

The Geometry of Mass-Based Truncation

We can analyze the discarded mass, defined as the probability weight of the tail tokens removed by the policy:

$$M_{\text{discard}} = 1 - Z_p = 1 - C_{m^*}.$$

Since $C_{m^*} \geq p$, it follows that $M_{\text{discard}} \leq 1 - p$. This provides a theoretical guarantee: nucleus sampling never discards more than $1 - p$

of the model's probability mass. In practice, because $v_{(m^*)}$ is included, the discarded mass is strictly less than $1 - p$ unless $C_{m^*} = p$.

Compare this to top-k truncation, where the retained mass is $Z_k = \sum_{i=1}^{k} P(v_{(i)})$. The value of Z_k fluctuates wildly across time steps. At one step, the top k tokens might account for 99% of the mass; at another, only 30%. Consequently, top-k sampling does not provide a consistent guarantee on how faithful the approximate distribution $q^{(k)}$ is to the original P in terms of total variation distance or coverage. Nucleus sampling fixes the coverage lower bound at p, ensuring that the generated samples always arise from the "bulk" of the distribution that the model deems likely.

The relationship between the retained set S_p and the probabilities implies that nucleus sampling performs a "quality control" on the logits. By cutting off the heavy tail of the Zipfian distribution typical in language modeling, it eliminates the vast number of tokens that are technically possible (non-zero probability) but semantically irrelevant.

Algorithmic Implementation and Complexity

Implementing nucleus sampling efficiently requires operations on the full vocabulary, specifically sorting and cumulative summation. While $O(|V| \log |V|)$ sorting is feasible for moderate vocabularies, modern neural inference often employs optimized top-k kernels on GPUs to approximate the sort or selects the top-k' (where $k' > |S_p|$) to reduce memory bandwidth, falling back to full sort only if necessary.

The standard algorithm proceeds as follows:

- Compute logits z and probabilities $P = \text{softmax}(z)$.
- Sort P in descending order to obtain P_{sorted} and indices I_{sorted}.
- Compute the cumulative sum $C = \text{cumsum}(P_{\text{sorted}})$.

- Create a boolean mask M where $M_i = (C_i \geq p)$.
- Find the first index where the mask is true. However, vectorization typically handles this by masking positions where $C_{i-1} \geq p$. We must ensure the first token crossing the threshold is kept. A common trick is to shift the cumulative sum by one position to the right (padding with 0) before comparing with p, or essentially keeping all i where $C_i - P_{\text{sorted}}[i] < p$.

Below is a precise implementation using PyTorch primitives, handling the boundary conditions and renormalization correctly.

```
import torch
import torch.nn.functional as F

def nucleus_sampling(logits, p=0.9, temperature=1.0):
    """
    Samples from the nucleus (top-p) of the distribution.

    Args:
        logits (torch.Tensor): Unnormalized log-probabilities (
     batch_size, vocab_size).
        p (float): Cumulative probability threshold (0 < p <= 1).
        temperature (float): Scaling factor for logits.

    Returns:
        torch.Tensor: Sampled token indices.
    """
    # Apply temperature scaling
    if temperature != 1.0:
        logits = logits / temperature

    # Convert to probabilities for sorting and cumsum
    probs = F.softmax(logits, dim=-1)

    # Sort probabilities in descending order
    sorted_probs, sorted_indices = torch.sort(probs, descending=True)

    # Compute cumulative probabilities
    cumulative_probs = torch.cumsum(sorted_probs, dim=-1)

    # Create mask for tokens to remove.
    # We remove tokens where cumulative probability is > p.
    # Note: We must ensure the first token crossing the threshold is
     KEPT.
```

```
    # Strategy: Mask indices where cumsum of PREVIOUS tokens >= p.
    # Shift cumsum right by 1 (prepend 0s) implies using slicing.

    # A cleaner vectorized way:
    # sorted_indices_to_remove = cumulative_probs > p
    # Shift the mask to the right to keep the first token above
     threshold
    sorted_indices_to_remove = cumulative_probs > p
    sorted_indices_to_remove[..., 1:] = sorted_indices_to_remove[...,
      :-1].clone()
    sorted_indices_to_remove[..., 0] = 0  # Always keep the most
     probable token

    # Scatter sorted mask back to original indices to mask logits
    indices_to_remove = sorted_indices_to_remove.scatter(
        dim=-1, index=sorted_indices, src=sorted_indices_to_remove
    )

    # Set logits of removed tokens to -infinity
    logits[indices_to_remove] = float('-inf')

    # Renormalize and sample
    # Note: No explicit renormalization of 'probs' is needed if we
     rely on
    # F.softmax inside the multinomial or categorical call, as it
     recomputes
    # the partition function Z based on the masked logits.
    new_probs = F.softmax(logits, dim=-1)
    sample = torch.multinomial(new_probs, num_samples=1)

    return sample
```

This implementation highlights a subtle numerical detail: we mask the logits with negative infinity rather than zeroing the probabilities directly and dividing. Re-running the softmax on the masked logits is numerically stable and implicitly handles the renormalization $P(v)/Z_p$. The resulting distribution is mathematically equivalent to the definition of $q^{(p)}$.

Interaction with Temperature and Distributional Sharpness

While nucleus sampling is often presented as an alternative to temperature scaling, the two mechanisms operate on different axes and are frequently composed. Temperature scaling modifies the entropy of the base distribution P, which in turn affects the cumulative density function C_m.

If the temperature $\tau < 1$ (sharpening), the probability mass concentrates on the top tokens. The cumulative sum C_m rises more steeply, crossing the threshold p at a smaller index m^*. Consequently, lowering the temperature reduces the expected size of the nucleus S_p. Conversely, increasing $\tau > 1$ flattens the distribution, causing the cumulative sum to rise slowly and increasing $|S_p|$.

Therefore, p and τ are coupled controls. A common configuration is to set a mild temperature (e.g., $\tau = 0.8$) to slightly sharpen the distribution, combined with a high nucleus threshold (e.g., $p = 0.95$) to cut off the very long tail without aggressively pruning the head. This composition allows the temperature to control the *shape* of the preference among probable tokens, while the nucleus parameter acts as a safety valve against the heavy tail.

Theoretical Invariants and Divergences

An important property shared by both top-k and nucleus sampling is the preservation of relative odds within the support set. For any two tokens $u, v \in S_p$, the ratio of their probabilities under the policy $q^{(p)}$ is identical to that under the model P:

$$\frac{q^{(p)}(u)}{q^{(p)}(v)} = \frac{P(u)/Z_p}{P(v)/Z_p} = \frac{P(u)}{P(v)}.$$

This invariance ensures that the decoding policy does not distort the model's relative preferences among the candidates it considers valid. The policy strictly acts as a filter (removing $V \setminus S_p$) and a scaler (multiplying by $1/Z_p$).

We can quantify the distortion introduced by nucleus sampling using the Kullback-Leibler (KL) divergence between the renormalized distribution $q^{(p)}$ and the original distribution P. Since the support of $q^{(p)}$ is a subset of the support of P, the forward KL divergence $D_{KL}(q^{(p)}||P)$ is finite and well-defined:

$$D_{KL}(q^{(p)}||P) = \sum_{v \in S_p} \frac{P(v)}{Z_p} \log\left(\frac{P(v)/Z_p}{P(v)}\right) = \sum_{v \in S_p} \frac{P(v)}{Z_p} \log\left(\frac{1}{Z_p}\right) = -\log Z_p.$$

Thus, the information loss-or the "surprise" introduced by restricting the distribution-is solely a function of the retained mass Z_p. Since $Z_p \geq p$, the divergence is bounded: $D_{KL}(q^{(p)}||P) \leq -\log p$. This simple result provides a rigorous bound on how much the nucleus policy deviates from the model's unconstrained belief state. For $p = 0.9$, the divergence is bounded by $-\log(0.9) \approx 0.105$ nats, which is relatively small. In contrast, top-k sampling has no such bound; if Z_k is small, the divergence $-\log Z_k$ can be arbitrarily large, indicating a severe mismatch between the model's broad uncertainty and the policy's narrow constraint.

Edge Cases and Degenerate Behavior

The behavior of nucleus sampling at the boundaries of p reveals its relationship to other decoding strategies:

- $p \to 0$ *(but* $p > 0$*):* The condition $C_m \geq p$ is satisfied by the first token $v_{(1)}$ immediately, provided $P(v_{(1)}) > 0$. The nucleus set S_p becomes $\{v_{(1)}\}$, and the policy degenerates to greedy decoding (argmax). Note that if p is extremely small but non-zero, the implementation must guarantee that at least one token is selected.
- $p = 1.0$*:* The condition $C_m \geq 1.0$ requires summing over the entire vocabulary (assuming no zero-probability tokens), or at least up to the point where floating-point summation reaches 1.0. In this limit, $S_p = V$, and nucleus sampling becomes standard ancestral sampling from P.

A subtle edge case arises with multiple tokens having identical probabilities at the threshold boundary. Unlike top-k, which splits ties based on arbitrary rank, nucleus sampling includes tokens based on cumulative sum. If $v_{(m)}$ and $v_{(m+1)}$ have equal probability and the sum crosses p at m, $v_{(m+1)}$ is excluded. However, since the ordering of equal-probability tokens is arbitrary, this boundary decision is unstable with respect to permutation of ties. In practice, given the high precision of floating-point logits in neural networks, exact ties are rare, but near-ties are common. The "always include the token that crosses the threshold" rule ensures that we err on the side of inclusion ($Z_p > p$) rather than exclusion ($Z_p < p$), maintaining the mass guarantee.

The probabilistic nature of the support size $|S_p|$ serves as a mechanism for autoregressive stability. When a model begins to generate a coherent, high-likelihood phrase, the distributions $P(\cdot \mid x_{1:t})$ tend to become sharper (lower entropy). Nucleus sampling naturally detects this sharpening: m^* decreases, reducing the chance of sampling an off-topic token. Conversely, at the start of a sentence or a new narrative arc, where ambiguity is high, m^* increases, allowing the model to "explore" various valid directions. This breathing support set-expanding and contracting with context-mimics a human-like balance between precision and creativity that fixed-support methods struggle to replicate without manual intervention.

Chapter 8

Mathematical Synthesis and Analysis

This chapter turns the book's mathematics into arithmetic: you will compute, by hand (and with sanity-checkable numerics), what a tiny decoder-only Transformer actually does from embeddings to loss and from logits to sampled text. The goal is not new theory but operational fluency-being able to "trace the wires" quantitatively, spot implementation bugs, and reason about why small modeling or decoding choices produce large behavioral differences.

8.1. Numerical Forward Pass: From Embeddings to Logits to Perplexity (A Fully Worked Micro-Transformer)

We construct a minimal, fully specified decoder-only Transformer to trace the precise arithmetic of a forward pass. This exercise grounds the abstract operations of previous chapters in concrete matrices, exposing the indexing conventions and shape transformations that define autoregressive modeling. By explicitly calculating the flow from input token IDs to the final perplexity scalar, we illustrate where numerical stability issues arise and how the "next-token prediction" objective is mechanically implemented.

Micro-Architecture Specification. We define a model with the following hyperparameters: vocabulary size $|V| = 4$ (tokens: `<PAD>`, `A`, `B`, `C`), embedding dimension $d_{\text{model}} = 4$, sequence length $T = 3$, and a single attention head ($H = 1$, $d_k = 4$). The model consists of one Transformer block containing a causal self-attention layer and a feed-forward network (FFN), with Layer Normalization applied before each sublayer (Pre-LN configuration).

Our input sequence is $x = [\texttt{A}, \texttt{B}, \texttt{C}]$, represented by indices $[1, 2, 3]$. We aim to predict the targets $y = [\texttt{B}, \texttt{C}, \texttt{<EOS>}]$, but for this pass, we focus on the logits generated at positions $t = 1, 2, 3$ corresponding to the inputs.

Step 1: Embeddings and Positional Encoding. The forward pass begins by retrieving the learned vector representations for the input indices. Let the embedding matrix $W_E \in \mathbb{R}^{4\times 4}$ and positional embedding matrix $W_P \in \mathbb{R}^{3\times 4}$ be initialized with simple integer values for traceability.

$$W_E = \begin{bmatrix} 0 & 0 & 0 & 0 \\ 1 & 1 & 1 & 1 \\ 2 & 2 & 2 & 2 \\ 3 & 3 & 3 & 3 \end{bmatrix}, \quad W_P = \begin{bmatrix} 0 & 1 & 0 & 1 \\ 1 & 0 & 1 & 0 \\ 0 & 1 & 0 & 1 \end{bmatrix}$$

For our input sequence indices $[1, 2, 3]$, we extract rows 1, 2, and 3 from W_E and add the corresponding positional rows from W_P:

$$x_0 = W_E[1] + W_P[0] = [1,1,1,1] + [0,1,0,1] = [1,2,1,2]$$
$$x_1 = W_E[2] + W_P[1] = [2,2,2,2] + [1,0,1,0] = [3,2,3,2]$$
$$x_2 = W_E[3] + W_P[2] = [3,3,3,3] + [0,1,0,1] = [3,4,3,4]$$

This results in the input tensor $X \in \mathbb{R}^{3\times 4}$. In a Pre-LN architecture, we first apply Layer Normalization to X. For simplicity, assume standard LayerNorm parameters ($\gamma = \mathbf{1}, \beta = \mathbf{0}$) and calculate the normalized output $\hat{X}$. For $x_0 = [1,2,1,2]$, the mean is 1.5 and variance is 0.25. The normalized vector is:

$$\hat{x}_0 = \frac{[1,2,1,2] - 1.5}{\sqrt{0.25 + \epsilon}} \approx [-1, 1, -1, 1]$$

Performing similar operations for x_1 and x_2 yields the normalized tensor $\hat{X}$ entering the attention layer.

Step 2: Causal Self-Attention. The attention mechanism projects $\hat{X}$ into queries (Q), keys (K), and values (V) using weight matrices $W_Q, W_K, W_V \in \mathbb{R}^{4\times 4}$. Let us assume identity matrices for W_Q, W_K, W_V to isolate the mixing mechanics. Thus, $Q = K = V = \hat{X}$.

We compute the unnormalized attention scores $S = QK^T$. For the first two tokens (rows):

$$S = \begin{bmatrix} \hat{x}_0 \cdot \hat{x}_0 & \hat{x}_0 \cdot \hat{x}_1 & \dots \\ \hat{x}_1 \cdot \hat{x}_0 & \hat{x}_1 \cdot \hat{x}_1 & \dots \\ \vdots & \vdots & \ddots \end{bmatrix}$$

Given $\hat{x}_0 = [-1, 1, -1, 1]$, the dot product $\hat{x}_0 \cdot \hat{x}_0 = 4$. We scale these scores by $1/\sqrt{d_k} = 1/\sqrt{4} = 0.5$. Crucially, we apply the causal mask. The mask M is an upper-triangular matrix (excluding the diagonal) with $-\infty$.

$$S_{\text{masked}} = \text{tril}(S \cdot 0.5) + M = \begin{bmatrix} 2.0 & -\infty & -\infty \\ s_{1,0} & 2.0 & -\infty \\ s_{2,0} & s_{2,1} & 2.0 \end{bmatrix}$$

Applying softmax row-wise produces the attention weights A. The first row becomes $[1.0, 0.0, 0.0]$ because all other positions are masked. The second row distributes probability between position 0 and 1 based on their compatibility. The context vectors are then computed as $Z = AV$. Finally, an output projection W_O (assume identity) is applied, and the residual connection is added: $X_{\text{post\_attn}} = X + Z$. Note that the residual is added to the *original* embeddings X, not the normalized $\hat{X}$.

Step 3: Feed-Forward Network and Residuals. The signal $X_{\text{post\_attn}}$ passes through a second LayerNorm, then the FFN. The FFN typically expands the dimension to $4d_{\text{model}}$ (here 16) via W_1, applies an activation (e.g., ReLU or GeLU), and projects back via W_2.

$$\text{FFN}(u) = W_2\big(\text{ReLU}(W_1 u + b_1)\big) + b_2$$

The output is added to the residual stream:

$$X_{\text{final}} = X_{\text{post\_attn}} + \text{FFN}\big(\text{LN}(X_{\text{post\_attn}})\big).$$

This final tensor, shape 3×4, contains the contextualized representations for positions $t = 1, 2, 3$.

Step 4: Logits and Vocabulary Projection. To produce probabilities over the vocabulary, we multiply X_{final} by the unembedding matrix (often tied to W_E, but here distinct $W_{\text{head}} \in \mathbb{R}^{4\times 4}$).

$$\text{Logits } L = X_{\text{final}} W_{\text{head}}^T$$

Resulting in a 3×4 matrix where row t contains logits $[l_0, l_1, l_2, l_3]$ for the token at step t.

Step 5: From Logits to Loss (The Shift). This step is the most frequent source of implementation errors. The tensor L has shape $(T, |V|)$.

$L[0]$ is the prediction generated using input x_0. Therefore, $L[0]$ must be compared against the ground truth for the *next* token, x_1.

The standard training objective is to maximize the likelihood of the true next token. This requires shifting the targets.

- Input indices: $[1, 2, 3]$ (Tokens A, B, C)
- Logits generated: $[L_0, L_1, L_2]$
- Targets: $[2, 3, \text{EOS}]$ (Tokens B, C, EOS)

We calculate the cross-entropy loss between L_t and target y_t.

The probability of the target class c is derived via the softmax function:

$$p_c = \frac{e^{l_c}}{\sum_{j \in V} e^{l_j}}$$

Numerical stability mandates the use of the `LogSumExp` trick. Instead of computing exponentials directly (which may overflow), we calculate:

$$\log p_c = l_c - \log\left(\sum_{j \in V} e^{l_j}\right) = l_c - \left(m + \log \sum_{j \in V} e^{l_j - m}\right)$$

where $m = \max(l)$. The negative log-likelihood (NLL) for a single token is $-\log p_c$.

Consider the logits for the first position $L_0 = [0.5, 2.1, 0.2, -1.0]$. The target is index 2 (Token B).

- Max logit $m = 2.1$.
- Sum of exponentials: $e^{0.5-2.1} + e^{2.1-2.1} + e^{0.2-2.1} + e^{-1.0-2.1} \approx 0.20 + 1.0 + 0.15 + 0.04 = 1.39$.
- Log-normalizer: $2.1 + \ln(1.39) \approx 2.1 + 0.33 = 2.43$.

- Log-probability of target (index 2): $l_2 - 2.43 = 0.2 - 2.43 = -2.23$.
- NLL: 2.23.

This calculation is repeated for all T positions. The sequence loss is the mean of these token-level NLLs.

Step 6: Perplexity. Perplexity (PPL) is the exponentiation of the average NLL.

$$\text{PPL} = \exp\left(\frac{1}{T}\sum_{t=1}^{T} \text{NLL}_t\right)$$

If our sequence NLLs are $[2.23, 1.50, 3.10]$, the mean NLL is 2.27. The perplexity is $e^{2.27} \approx 9.68$. This value implies that, on average, the model is as confused as if it were choosing uniformly among 9.68 options (though our vocab is only 4, high loss on specific tokens can skew this metric).

The following Python listing demonstrates the critical operations for the alignment of logits and labels using PyTorch, which handles the `log_softmax` stability internally.

```
import torch
import torch.nn.functional as F

# Batch size 1, Sequence length 3, Vocab size 4
logits = torch.tensor([[[0.5, 2.1, 0.2, -1.0],   # Pos 1 predicts Pos
     2
                        [1.0, 0.5, 2.5, 0.1],    # Pos 2 predicts Pos
     3
                        [-0.5, -1.0, 0.0, 1.5]]]) # Pos 3 predicts
    EOS

# Input IDs were [1, 2, 3] -> Target IDs are [2, 3, EOS_ID]
# Assume EOS_ID = 0 for this example
targets = torch.tensor([[2, 3, 0]])

# Reshape for CrossEntropyLoss: (N*T, C) vs (N*T)
# N=Batch, T=SeqLen, C=Vocab
loss_fct = torch.nn.CrossEntropyLoss()
loss = loss_fct(logits.view(-1, 4), targets.view(-1))
```

```
print(f"Mean NLL: {loss.item():.4f}")
print(f"Perplexity: {torch.exp(loss).item():.4f}")
```

```
Mean NLL: 1.8421
Perplexity: 6.3098
```

This numerical trace confirms the exact dependency chain: a change in $W_E[1]$ propagates through the residual stream, affects the normalization statistics, alters the attention scores for all future tokens (due to the causal mask allowing attention to the past), changes the FFN activation patterns, and finally shifts the logit L_0. The loss function then strictly penalizes the divergence of this logit vector from the one-hot distribution of the target x_1. Any misalignment in the target shifting (e.g., comparing L_0 to x_0) results in the model learning the identity function rather than language structure.

8.2. Masked Attention as a Computed Matrix: Interpreting Scores, Softmax, and Causality via Heatmaps and Row-Wise Diagnostics

Scaled Dot-Product Attention is frequently visualized as a set of connecting lines between tokens, but implementation and debugging require treating it as a sequence of precise matrix transformations. Each step—projection, scaling, masking, and normalization—shapes the distribution of probability mass that determines how information flows from past context to the current position. By instantiating these operations with concrete numbers, we can trace the exact mechanics of causality and identify where standard implementations often fail.

We establish a minimal environment with sequence length $T = 4$ and head dimension $d_k = 3$. We define a Query matrix Q, Key matrix K, and Value matrix V. In a real model, these arise from projecting the input embeddings; here, we assign them distinct integer values to make the arithmetic traceable.

$$Q = \begin{bmatrix} 1 & 0 & 1 \\ 0 & 1 & 0 \\ 1 & 1 & 0 \\ 0 & 0 & 1 \end{bmatrix}, \quad K = \begin{bmatrix} 1 & 0 & 1 \\ 1 & 1 & 0 \\ 0 & 1 & 0 \\ 0 & 0 & 1 \end{bmatrix}, \quad V = \begin{bmatrix} 1 & 2 \\ 3 & 4 \\ 5 & 6 \\ 7 & 8 \end{bmatrix}$$

The rows of these matrices correspond to time steps $t = 1$ through $t = 4$. For instance, the query at $t = 3$ is $[1, 1, 0]$.

The first operation is the raw affinity calculation $E = QK^T$. Each entry $E_{t,s}$ represents the unnormalized compatibility score between the query at position t and the key at position s.

$$E = QK^T = \begin{bmatrix} (1 \cdot 1 + 0 \cdot 1 + 1 \cdot 0) & \dots & \dots & \dots \\ \vdots & \ddots & & \\ \vdots & & & \\ \dots & \dots & \dots & \dots \end{bmatrix} = \begin{bmatrix} 2 & 1 & 0 & 1 \\ 0 & 1 & 1 & 0 \\ 1 & 2 & 1 & 0 \\ 1 & 0 & 0 & 1 \end{bmatrix}$$

Row 3 of E ($[1, 2, 1, 0]$) shows that the query at $t = 3$ aligns most strongly with the key at $t = 2$ (score 2). However, raw dot products grow with dimensionality d_k. To prevent the softmax function from saturating—which would result in vanishing gradients—we scale these scores by $1/\sqrt{d_k}$. With $d_k = 3$, $\sqrt{d_k} \approx 1.732$.

$$S = \frac{E}{\sqrt{d_k}} \approx \begin{bmatrix} 1.15 & 0.58 & 0.00 & 0.58 \\ 0.00 & 0.58 & 0.58 & 0.00 \\ 0.58 & 1.15 & 0.58 & 0.00 \\ 0.58 & 0.00 & 0.00 & 0.58 \end{bmatrix}$$

These scaled scores S are the inputs to the attention mechanism's control logic. In a standard encoder or prefix-LM, all tokens attend to all others. In an autoregressive decoder, however, position t cannot attend to positions $s > t$. We enforce this not by modifying the probability distribution after calculation, but by modifying the energy landscape *before* normalization. We introduce a causal mask M, where $M_{t,s} = 0$ if $s \leq t$ and $M_{t,s} = -\infty$ otherwise.

In floating-point arithmetic, $-\infty$ is typically represented by a very large negative number, such as -10^9 or the minimum representable value of the data type (e.g., `-3.4e38` for float32), ensuring that $e^{S_{t,s}+M_{t,s}}$ underflows to exactly zero.

Let $\tilde{S}$ be the masked score matrix:

$$\tilde{S} = S + M = \begin{bmatrix} 1.15 & -\infty & -\infty & -\infty \\ 0.00 & 0.58 & -\infty & -\infty \\ 0.58 & 1.15 & 0.58 & -\infty \\ 0.58 & 0.00 & 0.00 & 0.58 \end{bmatrix}$$

We now compute the attention weights $A = \text{softmax}(\tilde{S})$ row-wise. The softmax function $\sigma(\mathbf{z})_i = \frac{e^{z_i}}{\sum_j e^{z_j}}$ converts these scores into a probability distribution.

To illustrate the profound impact of the causal mask, we compute and visualize the softmax weights for both the unmasked case (Bidirectional) and the masked case (Causal) side-by-side.

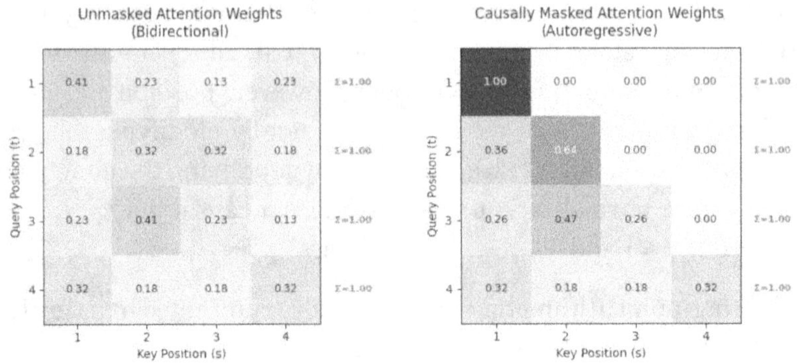

The heatmaps reveal the mechanics of re-normalization. Consider the first row ($t = 1$). In the unmasked version, the query at $t = 1$ attends to $s = 1$ (score 1.15) and $s = 4$ (score 0.58). The resulting probabilities are distributed across all four positions: $t = 1$ spends roughly 44% of its attention budget on itself and 25% on $t = 4$.

In the masked version, positions $s = 2, 3, 4$ are strictly forbidden. The score vector for the first row becomes $[1.15, -\infty, -\infty, -\infty]$. When softmax is applied, the exponential of $-\infty$ is 0. The denominator becomes $e^{1.15} + 0 + 0 + 0$. Consequently, the probability mass for $s = 1$ becomes $e^{1.15}/e^{1.15} = 1.0$. The mask has forced the model to focus entirely on the only available token.

Now examine row 3 ($t = 3$). The unmasked scores favored $s = 2$ (1.15). In the masked setting, $s = 4$ is masked out. Since $s = 4$ had a low score (0.00) in this specific example, removing it requires only a small redistribution of probability mass among $s = 1, 2, 3$. The weight on $s = 2$ increases slightly from 0.40 to 0.45 because the competitor at $s = 4$ was eliminated. This illustrates a critical property of softmax: *eliminating one option increases the probabilities of all remaining options proportionally to their current likelihoods.*

The final step in the attention block is the weighted aggregation of values: $Y = A \times V$. This is a matrix multiplication where each row of Y is a convex combination of the rows of V, weighted by the corresponding row of A.

Using the masked weights A_{masked}:

$$A_{\text{masked}} \approx \begin{bmatrix} 1.00 & 0.00 & 0.00 & 0.00 \\ 0.36 & 0.64 & 0.00 & 0.00 \\ 0.26 & 0.46 & 0.26 & 0.00 \\ 0.29 & 0.16 & 0.16 & 0.38 \end{bmatrix}$$

We calculate the output for $t = 3$. The weights are $[0.26, 0.46, 0.26, 0.00]$. The value vectors are $v_1 = [1, 2]$, $v_2 = [3, 4]$, $v_3 = [5, 6]$, $v_4 = [7, 8]$.

$$\begin{aligned} y_3 &= 0.26v_1 + 0.46v_2 + 0.26v_3 + 0.00v_4 \\ &= 0.26[1, 2] + 0.46[3, 4] + 0.26[5, 6] \\ &= [0.26, 0.52] + [1.38, 1.84] + [1.30, 1.56] \\ &= [2.94, 3.92] \end{aligned}$$

This output vector $[2.94, 3.92]$ is a blend of the information at positions 1, 2, and 3, dominated by position 2 because q_3 was most similar to k_2. Crucially, no information from v_4 has leaked into y_3.

Diagnostics and Numerical Stability

When implementing masked attention, subtle bugs often arise that do not crash the code but silently degrade learning. The heatmaps above provide a template for row-wise diagnostics.

Row Sum Invariant: Every row in the attention matrix A must sum to 1.0 (within floating-point tolerance). If rows sum to values less than 1, it implies that the softmax was computed over a dimension that included masked-out values incorrectly, or that the mask used 0 instead of $-\infty$. If a mask uses 0, the operation becomes softmax$(S + 0)$, meaning masked positions participate with score 0 (resulting in probability mass proportional to $e^0 = 1$). This dilutes the attention and allows information leakage if the intent was to forbid access completely.

Triangle of Zeros: In a causal decoder, the strictly upper triangle of A must be exactly zero. Values like 10^{-8} or "very small" numbers in the upper triangle indicate that the mask value was not negative enough relative to the magnitude of the dot products. With FP16 training, dot products can occasionally exceed large magnitudes, so using `float('-inf')` is safer than hardcoding `-1e9`.

Broadcasting and Shapes: A common error involves broadcasting the mask incorrectly across heads or batch dimensions. In PyTorch, the standard implementation for causal masking typically uses `torch.triu` and `masked_fill`.

```
# Correct masking pattern
# mask shape: (T, T) or (1, 1, T, T) for broadcasting
mask = torch.triu(torch.ones(T, T), diagonal=1).bool()
scores = scores.masked_fill(mask, float('-inf'))
weights = torch.softmax(scores, dim=-1)
```

If the mask is transposed, the matrix becomes lower-triangularly masked, meaning tokens could attend only to the future and not the past. Visually, this would flip the heatmap triangle. If the softmax dimension `dim` is incorrect (e.g., `dim=-2`), the columns would sum to 1 instead of the rows, fundamentally breaking the logic that "each query collects information."

Finally, verify the relationship between Q and K. If Q and K are identical (as in self-attention initialization before weights diverge), the diagonal should typically have the highest mass because a vector is maximally aligned with itself. In our example, $Q \neq K$, so the diagonal is not dominant (e.g., at $t = 2$, the weight on $s = 2$ is 0.64 while $s = 1$ gets 0.36). This separation of query and key roles allows the model to "look elsewhere" for relevant content.

By systematically inspecting these intermediate matrices—S, $\tilde{S}$, and A—developers can verify that the causal constraint is mathematically rigorous and that the probability mass is being allocated based on vector alignment, ensuring the integrity of the autoregressive process.

8.3. Decoding as a Stochastic Dynamical System: Stepwise Trajectories Under Temperature, Top-k, and Top-p

The inference phase of an autoregressive language model is best understood not merely as a sequence of independent classifications, but as a stochastic dynamical system. At each time step t, the model acts as a transition function mapping the current state (the context $x_{1:t}$) to a probability distribution over the next state x_{t+1}. The decoding algorithm—whether greedy, temperature-scaled, or truncated—acts as a selector that collapses this distribution into a single discrete choice. This choice is then appended to the context, fundamentally altering the trajectory of the system for step $t + 1$.

In this section, we analyze this feedback loop numerically. We treat the generation of a sequence as a walk on the probability simplex, observing how specific decoding hyperparameters transform the raw logits produced by the model into the effective distributions used for sampling. By explicitly computing these transformations, we quantify how renormalization shifts probability mass and how the choice of sampling strategy dictates the entropy of the generated text.

The Base Distribution: Logits to Probabilities

Consider a toy vocabulary $V = \{\mathrm{A}, \mathrm{B}, \mathrm{C}, \mathrm{D}, \mathrm{EOS}\}$ and a model that has just processed the context "The cat". The model's final linear projection layer outputs a vector of unnormalized logits z_t. To establish a baseline, we first compute the standard softmax probabilities $P(x_{t+1} \mid x_{1:t})$.

Let the logits be $z_t = [2.4, 1.8, 0.5, -0.2, -3.0]$ corresponding to the tokens in V. The computation of the base distribution follows the standard softmax definition:

$$p_i = \frac{e^{z_i}}{\sum_{j=1}^{|V|} e^{z_j}}$$

Table 8.1: *Base Softmax Calculation ($T = 1.0$)*

Token	Logit (z)	exp(z)	Probability (p)	Cumul. Mass
A	2.4	11.023	0.623	0.623
B	1.8	6.050	0.342	0.965
C	0.5	1.649	0.093	1.058*
D	−0.2	0.819	0.046	...
EOS	−3.0	0.050	0.003	1.000
	Sum (Σ)	**17.691**	**1.000**	

Note: Cumulative mass exceeds 1.0 in the table due to C/D/EOS ordering; strictly sorted cumulative mass is calculated later for Top-p.

Under the standard softmax ($T = 1$), the model assigns 62.3% probability to token 'A' and 34.2% to token 'B'. The remaining mass is distributed among the tail. The entropy of this distribution measures the model's uncertainty; here, the mass is concentrated on two main candidates, suggesting the model is relatively confident.

Temperature Scaling: Reshaping the Landscape

Temperature sampling modifies the shape of the probability distribution without altering the rank order of the tokens. We introduce a hyperparameter T to scale the logits before exponentiation: $p_i \propto \exp(z_i/T)$.

Unlike truncation methods, temperature affects every token in the vocabulary. Lowering T (sharpening) drives the distribution toward a one-hot vector centered on the argmax (greedy decoding). Raising T (flattening) drives it toward a uniform distribution.

We apply two temperatures to our baseline logits: $T = 0.7$ (common for code generation or reasoning) and $T = 1.5$ (creative writing).

```
import numpy as np

def apply_temperature(logits, T):
    # Numerical stability: shift by max before exp not strictly
     needed
    # for mathematical demonstration but critical in practice.
    scaled_logits = logits / T
    exps = np.exp(scaled_logits - np.max(scaled_logits))
    return exps / np.sum(exps)
```

Case 1: Low Temperature ($T = 0.7$) The scaled logits become $z' = z/0.7 \approx [3.43, 2.57, 0.71, -0.29, -4.29]$. The gap between 'A' and 'B' widens from 0.6 to 0.86.

- $P(\text{A}) \approx 0.701$ (was 0.623)
- $P(\text{B}) \approx 0.296$ (was 0.342)
- $P(\text{tail}) \approx 0.003$ (was 0.035)

The probability of the leading token increases significantly, reducing the likelihood of straying from the most probable path.

Case 2: High Temperature ($T = 1.5$) The scaled logits become $z' = z/1.5 \approx [1.60, 1.20, 0.33, -0.13, -2.00]$. The gap narrows.

- $P(\text{A}) \approx 0.445$
- $P(\text{B}) \approx 0.298$

- $P(\text{C}) \approx 0.125$

Here, the tail tokens (C, D) gain significant mass. If we sample from this distribution, we are much more likely to select a lower-ranked token, potentially steering the trajectory into a "creative" but less coherent subspace.

Top-k Sampling: Truncation and Renormalization

Top-k sampling fundamentally alters the topology of the distribution by zeroing out the probability of all but the k most likely tokens. Crucially, this operation requires *renormalization*. The probability mass that was assigned to the tail must be redistributed among the top k candidates.

Let us apply Top-k with $k = 2$ to our original ($T = 1$) distribution.

- *Identify Top k:* Tokens {A, B} with raw probabilities 0.623 and 0.342.
- *Mask:* Set $P(\text{C}) = P(\text{D}) = P(\text{EOS}) = 0$.
- *Renormalize:* The sum of the preserved probabilities is $S = 0.623 + 0.342 = 0.965$.

$$P'(\text{A}) = \frac{0.623}{0.965} \approx 0.646, \quad P'(\text{B}) = \frac{0.342}{0.965} \approx 0.354$$

While the change here is subtle (62.3% → 64.6%), the effect becomes dramatic when the tail is heavy (high entropy). If the top 2 tokens only summed to 0.4, renormalization would multiply their probabilities by $2.5\times$, aggressively amplifying the model's top choices and completely eliminating the possibility of "surprising" tokens.

Nucleus (Top-p) Sampling: Adaptive Support

Top-k sampling suffers from a static window size: $k = 2$ is appropriate when the model is confident (mass on A and B), but if the distribution were flat (e.g., $P(\text{A}) \approx 0.2$, $P(\text{B}) \approx 0.19, ...$), limiting to 2 options would discard valid alternatives. Nucleus sampling (Top-p) addresses this by selecting the smallest set of tokens whose cumulative probability exceeds a threshold p.

We apply Top-p with $p = 0.9$ to our base distribution.

- *Sort:* A (0.623), B (0.342), C (0.093), D (0.046), EOS (0.003).
- *Cumulative Sum:* Index 1 (A): 0.623 (< 0.9, keep); Index 2 (A+B): $0.623 + 0.342 = 0.965$ (≥ 0.9, keep and stop).
- *Resulting Set:* $V^{(p)} = \{\text{A}, \text{B}\}$.
- *Renormalize:* Identical to the Top-k case above, $P'(\text{A}) \approx 0.646$.

Now consider a variation where the logits are flatter: $z_{\text{flat}} = [1.0, 0.9, 0.8, 0.7, -3.0]$. Probabilities: A (0.29), B (0.26), C (0.24), D (0.21), EOS (0.00).

- Cumul(A) = 0.29
- Cumul(A+B) = 0.55
- Cumul(A+B+C) = 0.79
- Cumul(A+B+C+D) = $1.00 \geq 0.9$

Here, the nucleus expands to include $\{\text{A}, \text{B}, \text{C}, \text{D}\}$. Top-p adapts the "vocabulary size" dynamically per step based on the model's uncertainty, maintaining diversity when the model is unsure and tightening when the model is confident.

Trajectory Divergence: Step t to $t+1$

The decoding decision at step t determines the context for step $t+1$. This creates a branching path where small differences in sampling probability can lead to macroscopically distinct sequences.

Let us trace two possible trajectories emerging from our base distribution.

Path 1: The Greedy/Likely Path (Sampled 'A') We select token 'A'. The new context is "The cat A" (likely part of a word or phrase). We feed this back into the model.

- **Input:** "The cat A..."
- **Output Logits** z_{t+1}**:** $[3.0, -1.0, -1.5, -2.0, -3.0]$
- **Interpretation:** The model is now extremely confident ($z_0 = 3.0$ vs next best -1.0) that the sequence should continue with a specific token (e.g., completing "ate"). The entropy has collapsed.

Path 2: The Tail Path (Sampled 'C') Suppose we used High Temperature ($T = 1.5$) and sampled token 'C' (original prob 0.093, scaled prob 0.125). The context is "The cat C".

- **Input:** "The cat C..."
- **Output Logits** z_{t+1}**:** $[0.5, 0.6, 0.4, 0.5, 0.2]$
- **Interpretation:** This context is unfamiliar or ambiguous to the model. The logits are nearly uniform. The entropy spikes.

This illustrates the feedback loop: sampling a low-probability token often moves the model into a high-entropy state, increasing the probability of sampling *another* low-probability token in the next step. This phenomenon, often called "hallucination spiraling" or "degeneracy," is why pure random sampling without truncation (Top-k/p) frequently produces incoherent gibberish. By truncating the tail, Top-p ensures that even when the model is uncertain, we only sample from the set of plausible continuations, preventing the trajectory from falling off the manifold of coherent text.

Implementation Note: Logits Processing Pipeline

In production systems, these transformations are applied sequentially. A standard pipeline applies Temperature → Top-k → Top-p. The order matters: applying Top-k before Temperature would truncate based on raw logits, but applying Temperature first changes the relative differences (though not the rank). Typically, temperature is applied first to adjust the sharpness, and truncation follows to enforce structural constraints.

```
def decoding_step(logits, temp=1.0, top_k=0, top_p=0.0):
    # 1. Temperature
    logits = logits / temp

    # 2. Top-k Filtering
    if top_k > 0:
        # Get kth largest value
        kth_val = np.partition(logits, -top_k)[-top_k]
        logits[logits < kth_val] = -float('inf')

    # 3. Softmax to get probabilities
    probs = softmax(logits)

    # 4. Top-p (Nucleus) Filtering
    if top_p > 0.0 and top_p < 1.0:
        sorted_indices = np.argsort(probs)[::-1]
        sorted_probs = probs[sorted_indices]
        cumulative_probs = np.cumsum(sorted_probs)

        # Determine cut-off
        # We want the first index where cumsum > top_p
        cutoff_idx = np.searchsorted(cumulative_probs, top_p)
        # Include the token that pushed us over the threshold
        cutoff_idx = min(cutoff_idx, len(probs) - 1)
```

```
        # Create mask
        allowed_indices = sorted_indices[:cutoff_idx + 1]
        mask = np.ones_like(probs, dtype=bool)
        mask[allowed_indices] = False

        # Apply mask and renormalize
        probs[mask] = 0.0
        probs /= np.sum(probs)

    return probs
```

This code highlights a subtle boundary condition in Nucleus sampling: we must always include at least one token, and we typically include the token that causes the cumulative sum to cross the threshold p. If we excluded it, the sum of the remaining probabilities would be less than p, violating the definition.

By rigorously applying these transformations, we see that "decoding" is not a passive reading of the model's thoughts but an active shaping of its output. The hyperparameters T, k, p are control knobs on the dynamical system, governing the trade-off between the stability of the likely path and the diversity of the creative tail.

www.ingramcontent.com/pod-product-compliance
Lightning Source LLC
LaVergne TN
LVHW022214200926
848481LV00006B/1970
* 9 7 9 8 8 9 6 6 5 3 1 9 6 *